T0392412

VOLUME ONE HUNDRED AND NINETY FIVE

METHODS IN CELL BIOLOGY

Flow Cytometry in Immuno-Oncology

SERIES EDITOR

Lorenzo Galluzzi
Fox Chase Cancer Center,
Philadelphia, PA, United States

VOLUME ONE HUNDRED AND NINETY FIVE

METHODS IN CELL BIOLOGY

Flow Cytometry in Immuno-Oncology

Edited by

MARCELLO PINTI
University of Modena and Reggio Emilia, Modena, Italy

ANDREA COSSARIZZA
University of Modena and Reggio Emilia, Modena; Istituto Nazionale per le Ricerche Cardiovascolari, Bologna, Italy

Academic Press is an imprint of Elsevier
125 London Wall, London EC2Y 5AS, United Kingdom
525 B Street, Suite 1650, San Diego, CA 92101, United States
50 Hampshire Street, 5th Floor, Cambridge, MA 02139, United States

First edition 2025

ISBN: 978-0-323-89883-6
ISSN: 0091-679X

For information on all Academic Press publications
visit our website at https://www.elsevier.com/books-and-journals

Publisher: Zoe Kruze
Editorial Project Manager: Devwart Chauhan
Production Project Manager: A. Maria Shalini
Cover Designer: Arumugam Kothandan

Typeset by STRAIVE, India

Contents

Contributors

Andrea Antonuzzo
Unit of Medical Oncology 1, AOU Pisana, Pisa, Italy

Beatrice Aramini
Division of Thoracic Surgery, Department of Medical and Surgical Sciences—DIMEC of the Alma Mater Studiorum—University of Bologna and G.B. Morgagni—L. Pierantoni Hospital, Forlì, Italy

Brian O. Bachmann
Department of Chemistry; Department of Pharmacology, Vanderbilt University; Vanderbilt Institute of Chemical Biology, Nashville, TN, United States

Joseph A. Balsamo
Department of Pharmacology, Vanderbilt University, Nashville, TN, United States

Rebecca Borella
Department of Medical and Surgical Sciences for Children & Adults, University of Modena and Reggio Emilia, Modena, Italy

Jolene A. Bradford
Thermo Fisher Scientific, Fort Collins, CO, United States

Giuditta Comolli
Department of Microbiology and Virology and Laboratory of Biochemistry-Biotechnology and Advanced Diagnostics, IRCCS San Matteo Foundation, Pavia, Italy

Andrea Cossarizza
Department of Medical and Surgical Sciences for Children & Adults, University of Modena and Reggio Emilia, Modena; Istituto Nazionale per le Ricerche Cardiovascolari, Bologna, Italy

Marco Danova
Unit of Internal Medicine and Medical Oncology, Vigevano Civic Hospital, Pavia; LIUC University, Castellanza, Varese, Italy

Sara De Biasi
Department of Medical and Surgical Sciences for Children & Adults, University of Modena and Reggio Emilia, Modena, Italy

Massimo Dominici
Department of Medical and Surgical Sciences for Children & Adults, University of Modena and Reggio Emilia, Modena, Italy

Wendy N. Erber
Department of Haematology, PathWest Laboratory Medicine, Nedlands; School of Biomedical Sciences, The University of Western Australia, Perth, WA, Australia

Kathy A. Fuller
School of Biomedical Sciences, The University of Western Australia, Perth, WA, Australia

Lara Gibellini
Department of Medical and Surgical Sciences for Children & Adults, University of Modena and Reggio Emilia, Modena, Italy

Madeline J. Hayes
Department of Cell and Developmental Biology, Vanderbilt University; Department of Pathology, Microbiology and Immunology, Vanderbilt University Medical Center, Nashville, TN, United States

Henry Y.L. Hui
School of Biomedical Sciences, The University of Western Australia, Perth, WA, Australia

Jonathan M. Irish
Vanderbilt Institute of Chemical Biology; Vanderbilt Chemical and Physical Biology Program; Department of Cell and Developmental Biology, Vanderbilt University; Department of Pathology, Microbiology and Immunology; Vanderbilt-Ingram Cancer Center, Vanderbilt University Medical Center, Nashville, TN, United States

J. Jacobberger
Professor Emeritus, Case Western Reserve University, Cleveland, OH, United States

Jordi Juncà
Functional Cytomics Lab, Germans Trias i Pujol Research Institute (IGTP), ICO-Hospital Germans Trias i Pujol, Universitat Autònoma de Barcelona, Badalona, Barcelona, Spain

Stephanie J. Lam
Department of Haematology, Fiona Stanley Hospital, Murdoch; Department of Haematology, PathWest Laboratory Medicine, Nedlands, WA, Australia

Domenico Lo Tartaro
Department of Medical and Surgical Sciences for Children & Adults, University of Modena and Reggio Emilia, Modena, Italy

Valentina Masciale
Department of Medical and Surgical Sciences for Children & Adults, University of Modena and Reggio Emilia, Modena, Italy

Giuliano Mazzini
Institute of Molecular Genetics IGM-CNR, Pavia, Italy

Annamaria Paolini
Department of Medical and Surgical Sciences for Children & Adults, University of Modena and Reggio Emilia, Modena, Italy

Jordi Petriz
Functional Cytomics Lab, Germans Trias i Pujol Research Institute (IGTP), ICO-Hospital Germans Trias i Pujol, Universitat Autònoma de Barcelona, Badalona, Barcelona; Department of Cellular Biology, Physiology and Immunology, Autonomous University of Barcelona (UAB), Cerdanyola del Vallès, Spain

Laura G. Rico
Functional Cytomics Lab, Germans Trias i Pujol Research Institute (IGTP), ICO-Hospital Germans Trias i Pujol, Universitat Autònoma de Barcelona, Badalona, Barcelona; Department of Cellular Biology, Physiology and Immunology, Autonomous University of Barcelona (UAB), Cerdanyola del Vallès, Spain

J. Paul Robinson
Distinguished Professor of Cytometry & Professor of Biomedical Engineering, Purdue University, West Lafayette, IN, United States

Roser Salvia
Functional Cytomics Lab, Germans Trias i Pujol Research Institute (IGTP), ICO-Hospital Germans Trias i Pujol, Universitat Autònoma de Barcelona, Badalona, Barcelona; Department of Cellular Biology, Physiology and Immunology, Autonomous University of Barcelona (UAB), Cerdanyola del Vallès, Spain

Andrea Sbrana
Department of Surgical, Medical and Molecular Pathology and Critical Care Area, University of Pisa; Service of Pneumo-Oncology, Unit of Pneumology, Pisa, Italy

Henry A.M. Schares
Department of Chemistry; Vanderbilt Chemical and Physical Biology Program, Vanderbilt University; Vanderbilt Institute of Chemical Biology, Nashville, TN, United States

Hannah L. Thirman
Vanderbilt Institute of Chemical Biology; Vanderbilt Chemical and Physical Biology Program, Vanderbilt University, Nashville, TN, United States

Michael D. Ward
Thermo Fisher Scientific, Fort Collins, CO, United States

CHAPTER ONE

The evolution of flow cytometry with respect to cancer

J. Paul Robinson[a,*] and J. Jacobberger[b]

[a]Distinguished Professor of Cytometry & Professor of Biomedical Engineering, Purdue University, West Lafayette, IN, United States
[b]Professor Emeritus, Case Western Reserve University, Cleveland, OH, United States
*Corresponding author: e-mail address: jpr@cyto.purdue.edu

Contents

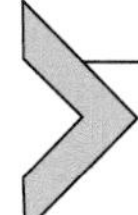

1. Background to analytical cancer detection, 1930–1985

In 1963 Louis Kamentsky developed a flow cytometer performed ultraviolet absorption measurements on epidermoid cancer cells (Kamentsky, Derman, & Melamed, 1963). The concept of evaluating the absorption spectrum of cells was not new, but Kamentsky's implementation in a flow cytometer certainly was. Evaluating the spectrophotometric properties of cells originated in the work of Caspersson (Caspersson, 1941), who defined the presence of ribonucleic acid by measuring the UV absorption of purine and pyrimidine bases. This was the first recognition that nucleic acids were actually components of all cells. Hard as it is to imagine, at that time the current understanding was that nucleic acids were more related to cellular debris or waste product. But it was not until 1950 that Caspersson published his definitive paper entitled "Cell Growth and Cell Function" that described nucleic acid and protein metabolism during normal and abnormal growth.

Methods in Cell Biology, Volume 195
ISSN 0091-679X
https://doi.org/10.1016/bs.mcb.2024.11.005

Again, these studies were performed using a cadmium spark for a UV light and electronic circuits for detection of absorption signals (Caspersson, 1950). Ironically, Oswald T. Avery had shown in 1944 that in fact DNA was the carrier of genetic information (Avery, McCarty, & MacLeod, 1944). But, as beautifully described in an article published in 2002 by Peter Reichard, a contemporary of Caspersson's, Avery's work was not considered Nobel Prize–worthy by the Nobel Committee (Reichard, 2002). Of course, we know that the definitive work on the structure of DNA considered by the Nobel committee was that of Watson and Crick (1953). What is clear is that the development of cell-analysis technologies mirrored simultaneous discoveries in fundamental biology, as shown in Fig. 1.

2. The driving force of clinical studies

Studies in uterine cancer should perhaps be given credit for the eventuality of the use of flow cytometry in cancer studies. Consider the landmark papers by Papanicolaou and Traut (Papanicolaou & Traut, 1941; Papanicolaou & Traut, 1943), which were the driving force for the development of the Cytoanalyzer instrument by Mellors and Silver in 1952 (Mellors & Silver, 1951); the principles of analysis were tested by Papanicolaou in 1952 (Mellors, Keane, & Papanicolaou, 1952). The fundamental method that Papanicolaou used was microphotometry of cells to obtain relative measures of the amount of DNA in individual cells. The concept here was that by measuring and comparing the total nucleic-acid content in cancer cells and in normal cells, abnormalities could be identified. These authors measured cellular nucleic acid via the UV spectrum, a difficult task that required very complex instrumentation; the spectra were eventually recorded onto photographic film! The Cytoanalyzer capitalized on this approach in an attempt to build an instrument that could reproducibly identify cancer cells in a clinical environment (Anonomous., 1957; Courtney et al., 1960; Diacumakos, Day, & Kopac, 1962; Tolles, 1955). This was effectively some of the earliest attempts to use some kind of electronic device for automated analysis of cancer. Fig. 1 provides a timeline for the integration of technologies over a century or more, although there are clearly combinations of technology and molecular discovery.

By the early 1960s, technologies were investigated with the focus on quantification and automation of pathological cancer diagnosis. There were at the time only three general options—absorbance of UV as defined above, imaging technologies, and light scatter or fluorescence. In the imaging

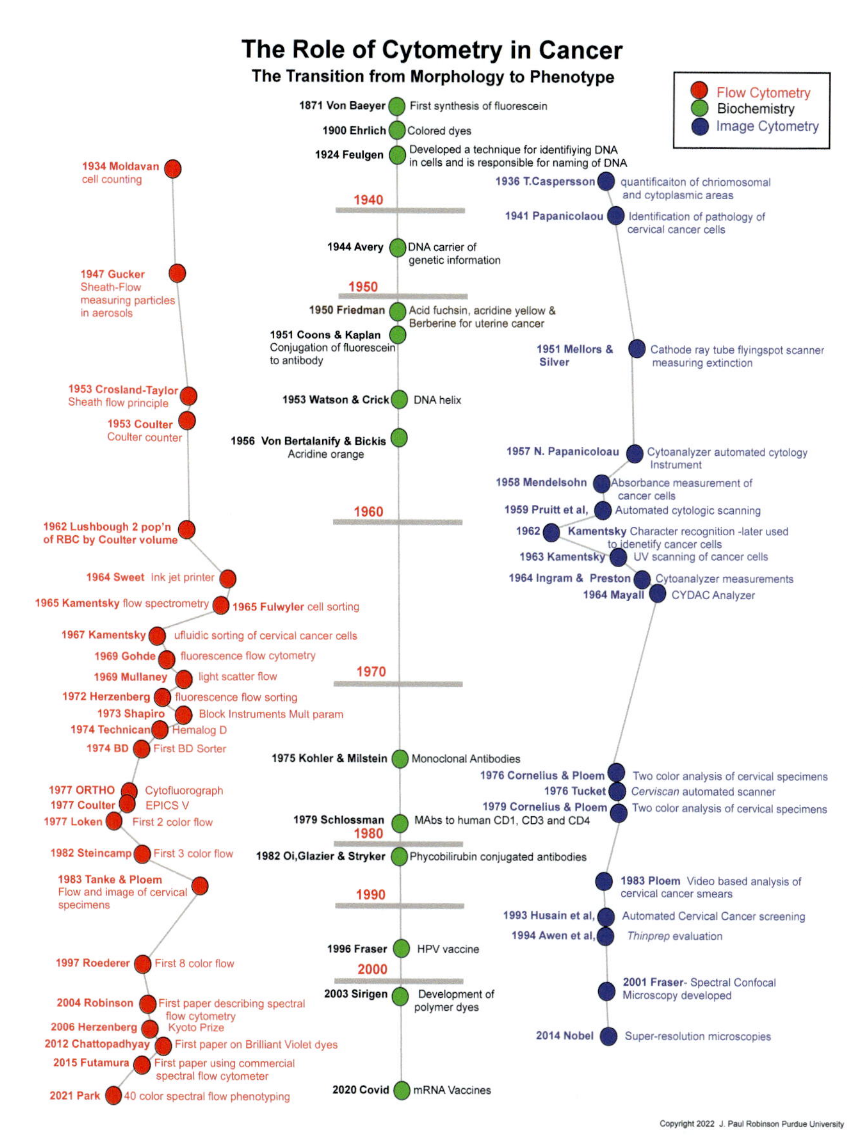

Fig. 1 Three components of cancer detection focused on cellular analysis or related areas are shown in this figure. On the left are key technical developments in cytometry, in the center are the related biochemistry discoveries and on the right image based advances that all impacted the pathway of discovery in cytometry. (Image is from PUCL Website and used with permission).

world, while most groups working on cancer were able to do regular microscopy, Marylou Ingram had observed bilobed nuclei as apparent effects of radiation (Ingram & Barnes, 1951) and by the early 1960s, Ingram began by using semi-automated image processing for abnormal cell identification (Ingram & Preston, 1964); by the 1970s she had what was essentially a fully automated instrument operating (although it was rather slow) (Ingram & Preston, 1970).

As noted earlier, Lou Kamentsky was an engineer working at IBM on character recognition (Kamentsky, 1958, 1962), and his work resulted in a direct application to cancer-cell detection. He also worked on spectroscopic analysis technologies for identification of cancer cells through his colleague Myron Melamed (Kamentsky et al., 1963; Kamentsky & Melamed, 1967; Kamentsky, Melamed, & Derman, 1965; Melamed & Kamentsky, 1969; Melamed, Kamentsky, & Boyse, 1969). This even included an ability to physically sort cells, although at very slow speeds in a microfluidic system.

Another electronic system that initially used micro-spectroscopy was developed by Mort Mendelson (Mendelsohn, 1958), which morphed into a dual-wavelength spectrometer system (Mendelsohn & Richards, 1958) and eventually into the CYDAC Analyzer capable of using the fluorescence of gallocyanin chromalum and naphthol yellow S as part of its imaging system (Mendelsohn & Mayall, 1974). During the same time Garry Salzman attempted to use multiple angles of light scatter in flow cytometry to identify abnormal gynecologic specimens (Salzman, Crowell, Hansen, Ingram, & Mullaney, 1976); this was in the earliest days of flow cytometry when instruments were minimalist with respect to the number of available parameters.

By the mid 1970s there was a need for technologies for faster cell analysis with multiple parameters, which might enable identification of cancer cells. However, most of the approaches were based on either the absorption of UV light by nucleic acid, or on early methods of staining nucleic acid with broad-spectrum nuclear stains. Some attempts were made to analytically identify and sort (very slowly) cancer cells with the relatively new fluorescence dye acridine orange (Melamed, Adams, Traganos, Zimring, & Kamentsky, 1972; Melamed, Traganos, Sharpless, & Darzynkiewicz, 1976) using flow cytometry. By the late 1970s this work was expanding as DNA analysis was recognized as a major advantage of flow cytometry (Collste et al., 1980; Traganos, Darzynkiewicz, Sharpless, & Melamed, 1979). A relatively sophisticated flow-cytometric approach to cancer-cell detection was developed by Wheeless and colleagues (Robinson, Wheeless, Hespelt, & Wheeless, 1990) (Wheeless, Reeder, & O'Connell, 1990). Their approach used acridine orange at non-stoichiometric dye-binding conditions. The objective was not to quantify the fluorescence relative to DNA and RNA concentrations per cell, but rather to measure the size of the nucleus and the cytoplasm and calculate nuclear to cytoplasmic ratio, a common method for analysis of slides and smears in pathology (Vaickus & Tambouret, 2015). They did this using a home-built instrument that captured each individual fluorescence-intensity, time-based cell profile as the cell traversed a laser beam that was focused to its fraction delimited size (~5 μm for 488 nm, a technique known as slit-scanning (Wheeless & Patten, 1973a, 1973b). Wheeless and colleagues also captured images of cells in flow using computer-activated cameras to identify artifacts (Wheeless et al., 1979; Wheeless, Kay, Cambier, Cambier, & Patten, 1977).

In the early 1970s, Wolfgang Göhde and colleagues extensively utilized flow cytometry for cancer diagnostics. Göhde had previously published the first paper on fluorescence in flow cytometry (Dittrich & Göhde, 1969b) and was working with several groups on this new approach to cancer-cell analysis (Buchner et al., 1974; Buchner et al., 1974; Buchner, Gohde, Dittrich, & Barlogie, 1972; Buchner, Hiddemann, Schneider, Kamanabroo, & Gohde, 1973; Gohde, Dittrich, Zinser, & Prieshof, 1972; Reiffenstuhl, Severin, Dittrich, & Gohde, 1971; Schumann, Ehring, & Göhde, 1975; Schumann, Ehring, Gohde, & Dittrich, 1971; Schumann & Gohde, 1974).

One key concept that drove researchers and clinicians into flow cytometry was the obvious feature of flow cytometry as an emerging computer-driven field. Few technologies were capable of producing such high quality and high-resolution information in computer form that was a core feature of flow cytometry. Despite the limited computing capacity available, flow cytometry utilized it to the max. Microscopy was clearly the key driver for pathology; however, as noted earlier, automated imaging systems were not efficient or even commercially available. Thus, a key concept in the expansion of flow cytometry was the notion that imaging was inadequate to handle the issue of tumor-cell heterogeneity. Since flow cytometry could evaluate hundreds of thousands or potentially millions of cells even in the early days of the technology, it appeared to be the technology of choice for tumor evaluation. Conceptually, the separation of tumor from normal cells was a key function for flow cytometry. Thus, the notion that DNA content could be an effective tool became embedded in the flow

Table 1 Engineering concepts introduced into cell analysis.

Key individual (refs)	Components	Key technique introduced
Casperson (1)	Bulk cell analysis	Spectroscopy
Fulwyler (2)	Coulter volume Cell separation	Impedance (Coulter volume)
Göhde (3)	Microscopy Fluorescence Cell identification	Imaging
Herzenberg (4)	Single-cell analysis Fluorescence Cell identification	Flow cytometry Piezo-based sorting Fluorescence
Salzman (5)	Flow cytometry Light scatter	Cell identification by size and shape and content
Wheeless (6)	Flow cytometry	Slit-scanning flow
Basiji (7)	Flow cytometry	Imaging in flow cytometry
De Rosa (8)	Flow cytometry	High parameter phenotyping
Robinson (9)	Flow cytometry	Spectral flow cytometry
Tanner (10)	Flow cytometry	Mass cytometry
Bodenmiller (11)	Flow cytometry	Imaging mass cytometry

References for the above Table: (1) (Casperson, 1964); (2) (Fulwyler, 1965); (3) (Dittrich & Göhde, 1969b); (4) (Hulett, Bonner, Barrett, & Herzenberg, 1969); (5) (Salzman et al., 1975); (6) (Wheeless et al., 1979; Wheeless & Patten, 1973a); (7) (George et al., 2004); (8) (De Rosa, Brenchley, & Roederer, 2003); (9) (Robinson, 2004); (10) (Bandura et al., 2009); (11) (Giesen et al., 2014).

cytometry community. A significant literature was published on the notion that ploidy analysis could be a highly useful flow-cytometry feature; see Table 1 in (Barlogie et al., 1983).

At the same time, other phenotypic correlations that would enhance basic DNA analysis were under study. One such approach was the ability to distinguish RNA from DNA (Andreeff, Darzynkiewicz, Sharpless, Clarkson, & Melamed, 1980), for example addition of acridine orange flow cytometry to distinguish AML from ALL (Darzynkiewicz, Traganos, Sharpless, & Melamed, 1976). Using these techniques developed by Darzynkiewicz, it appeared to be possible to note a progressive increase in RNA content in several cell types, including lymphocytes, erythroid and myeloid precursors and plasma cells. This technique was used for discrimination of myeloblastic from lymphoblastic leukemia (Barlogie et al., 1983).

Because flow cytometry was able to produce such advanced multi-parameter data sets compared to other technologies, it was an obvious potential solution to the analysis of cancer cells. One key innovation was the ability to define an aspect of cellular size and structure using laser light scatter. Salzman characterized this principle of cell analysis (Salzman et al., 1975; Salzman, Crowell, Martin, et al., 1975) and in combination with Barlogie expanded this concept to fully describe all aspects of light scatter in flow cytometry (Salzman, Hiebert, Jett, & Bartholdi, 1980). A combination of light scatter and two fluorescence parameters enhanced separation of cancer cells from normal ones (Frost et al., 1979); however, these studies should be considered within the context of instruments with very limited capacity for multiparametric analysis, considering that the instrument used was one of the earliest commercially available (a FACS-1 from Becton Dickinson). Over the next 40 years, a number of new approaches were developed that allowed the area of cell analysis to expand its impact; see Table 1.

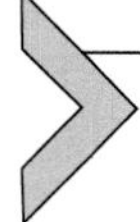

3. The role of DNA in cancer cell detection using flow cytometry

In the beginning, studies using fluorescence-based flow cytometry employed dyes because monoclonal antibodies had not yet been invented, poly-clonal sera were difficult to control, and probes that could be conjugated to antibodies were few (PubMed searches on flow cytometry AND "monoclonal antibody" show hits beginning in 1980, and by 1982 were equal in number to searches for lists of common dyes, which begin in 1977). Therefore, dyes that reported cellular nucleic acid, protein or lipid content, or membrane potential were studied most often. Of these dyes, those that were specific for DNA (Hoechst 33258, 33342), or reported DNA content after RNase treatment (ethidium, propidium), or could spectrally distinguish DNA and RNA (acridine orange) were the most investigated. The availability of DNA dyes fit well with the interest in measuring DNA content in the study of cancer. In a review of the emerging technology in an editorial in The Lancet in 1975, the editor stated:

"So far the most exciting development in flow-system analysis, has been fluorescence measurement of specifically stained DNA.... The advantages of this flow-system analysis over microspectrophotometry and autoradiography are the increased statistical reliability obtained by measuring 10^5–10^6 cells rather than 10^2–10^3 cells and the reduction in measurement time from days to minutes".

(Editorial, 1975)

It is interesting that one of the arguments for focusing on DNA with this new technology was a hypothesis clearly stated in the editorial:

"Data about the cell-cycle and kinetics of normal and malignant cells are important not only as basic information but also for practical therapy, because X rays and chemotherapeutic agents are more effective in certain phases of the cell-cycle than in others".

It was therefore somewhat inevitable that the primary application in the early implementation of flow cytometry focused on DNA analysis within the context of cancer and radiation biology. In cancer research, the questions were whether cancer cells could be distinguished from healthy cells by increases or decreases in the normal complement (C) of DNA—that is, a change in DNA-ploidy—and whether cancer-cell populations had more S-phase than healthy cells, since their rates of replication were higher. In radiation biology, an early question was whether exposure to ionizing radiation could be detected by changes in cell-cycle phase distributions of cell populations—cultured cells or tissues from experimental animals. The advent of monoclonal antibodies that could recognize incorporated bromo-deoxyuridine and other cytosine analogs improved the ability to measure altered cell-cycle kinetics because of the ability to follow a label through the cell cycle as a function of time. The ability to measure DNA content, bromo-deoxyuridine, and a mitotic marker (e.g., phospho-S10-histone H3) provided complete ability to perform stathmokinetic experiments. For a full review of the use of cytometry in radiation biology see (Wilson & Marples, 2007).

However, a problem in using flow cytometers faced by investigators of solid tissues was the requirement for single-cell suspensions. Tissue needed to be disassociated and the cells suspended. The drawbacks for this approach are that tissue architecture is lost and that reducing or eliminating the possibility of cell population recovery bias is significantly difficult. The literature is replete with papers and chapters on tissue dissociation (e.g., PubMed search on "single-cell suspension produces 996 items). In the limited experience (breast tumors, brain and prostate tissue) of one of us, methods using collagenase are preferred, but for any tissue, good advice is to empirically determine the conditions (Sigma-Aldrich Guide). Nevertheless, a PubMed search on (DNA content AND flow cytometry AND (tumor or tumor) returns 209 papers in the peak year of 1993. This represented 79% of the papers returned from a search on DNA Content AND flow cytometry.

Until 1983, the published work concerned studies of fresh tissue. In 1983, Hedley, Friedlander, Taylor, Rugg, & Musgrove (1983) published development of a technique that showed a good correlation between DNA histograms from fresh and fixed tissue from the same tumors. This single paper prompted a large number of studies throughout the world evaluating the vast storage of specimens in pathology labs. By 1990, 34% of DNA flow cytometry papers presented and/or discussed paraffin-embedded tissue. The driving force behind this effort was to correlate clinical outcome with features of DNA content analysis,. The S phase fraction is a measure of relative cycling rate and the idea is that the increased proliferation of tumor cells relative to healthy cells will be reflected in the measurement. DNA ploidy measures the number of stem lines that are not diploid in DNA content, which is potentially a measure of genome instability. Indeed, after 5 years of publications and over 100 additional studies it was evident that this approach had potential for clinical analysis (Hedley, 1989).

In 1993, a volume of Cytometry (now Cytometry Part A) presented the findings of a Consensus Conference on DNA content cytometry held the previous year. This issue presented a summary editorial, a technical guideline paper, and five consensus papers on the clinical utility of DNA content measurements of tumor cells in human specimens, including papers on bladder, breast, colorectal, haemopoietic, and prostate cancers. However, despite the generally positive consensus that DNA content variables were prognostic, technical problems were evident, conflicting reports existed, and complete validation remained to be done. The major technical problem may have been the absence of standardization, especially in acceptance/rejection criteria for specimens, data and analytical methodology.

In 1996, The ASCO Journal published the guidelines adapted by the American Society for Clinical Oncology for using tumor markers for clinical decisions for colorectal and breast cancers (Editorial, 1996). The guidelines state that the data were insufficient to recommend DNA index or DNA proliferation analysis. In 1996, papers were down (139) from the peak publication year, and by 2002 publications had fallen to 69 from the peak

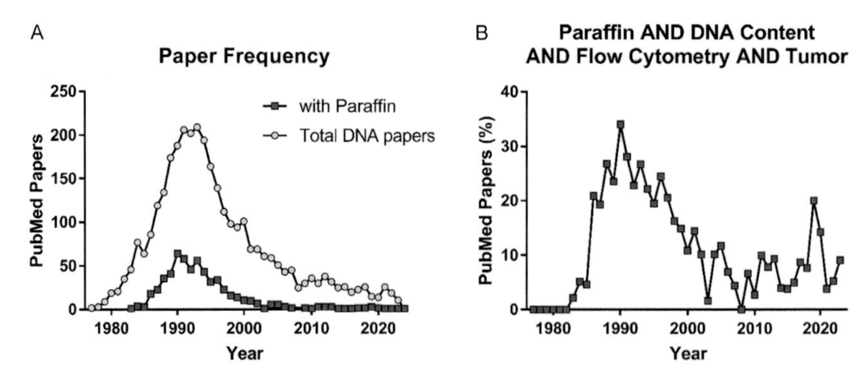

Fig. 2 Historical Frequency of Published Papers. A PubMed search of papers with criteria (A) "DNA Content" AND "flow cytometry" AND (Tumor OR Tumor) returns the data presented (O) using the left axis. (B) A search with criteria "DNA Content" AND "flow cytometry" AND (Tumor OR Tumor) AND paraffin returns data used to calculate the curve (□). Both plots show an interest in DNA content measurements of tumor cells that peaked in 1993. At the peak, approximately 27 % of papers published that mention DNA content also mention paraffin.

of 209 papers. The current average number of papers per year for years 2002–2024 is 33, whereas during the peak years (1984–2001), the average was 135 (see Fig. 2 for a graphical representation of publication frequencies). Within the cytometry community, some felt that the ASCO paper had "killed" the field. This was undoubtedly an over-reaction, and DNA content analysis remains part of some clinical laboratory practices. Bruce Bagwell has addressed the technical issues for node-negative breast cancer in three papers (Bagwell, 2004; Bagwell et al., 2001a, 2001b), and it is worth quoting the first paragraph in the 2004 paper. *"The combination of DNA ploidy and S-phase is one of the strongest general prognostic indicators for node-negative breast cancer (1). The costs associated with this prognostic power are relatively minimal; laboratory total processing time is usually less than 10 min and reagents are relatively inexpensive. The test can be standardized world-wide as long as strict guidelines are followed, and the test's result can be presented in a way that is easily understood by oncologists. DNA ploidy and S-phase prognostic strength can be further augmented by other well-known prognostic indicators, such as menopausal status and primary tumor size."* Current work utilizing DNA content analysis for gastrointestinal cancers can be assessed in the productive work of Peter Rabinovitch and colleagues—e.g., (Chao et al., 2008; Choi et al., 2015; Choi, Rabinovitch, Wang, & Westerhoff, 2015).

Early clinical flow cytometry focused on DNA because of the relationship of nuclear abnormalities in cancer tissue and radiation induced-damage and because fluorescent DNA binding dyes produce fluorescence intensities per cell that were proportional to DNA content. As noted above, absorption of UV light by nucleic acid, multi scatter, and fluorescence detection appeared at the time to be the key features that would reflect clinical potential. The ability to define cell cycle and quantify abnormal DNA content (DNA-aneuploidy) appeared to have great potential and a wave of research followed the paper by David Hedley (D. W. Hedley et al., 1983) that provided a method to obtain high-quality DNA content measurements from paraffin-embedded tissue, which is a historical tissue source within pathology laboratories. At the time it was considered that flow cytometry of DNA was likely to produce high-value for clinical cytometry, particularly with relation to cancer. Although the number of papers per year have diminished since then, the use of DNA content measurement of clinical tissue by cytometry continues to find use.

4. The relevance of conjugated antibodies

As previously noted, in the early days of flow cytometry only one or two parameters were available—typically a maximum of three or four in total—two scatter and one or two fluorescence signals. This meant that there was a restricted subset of cells that could be distinguished as there were few, if any, specific markers available. For decade, tracking any reagent in or attached to cells was achieved by a variety of fluorescent dyes that could be incorporated into cellular organelles or nucleic acid.

Ehrlich identified acidic and basic dyes to define acidophilic, eosinophilic, basophilic, and neutrophilic leukocytes in the 1880's, which enabled a study of the dynamics of ocular fluids and saw the introduction of fluorescein (Ehrlich & Lazarus, 1900). The predominant dyes available were saffron, methyl green, and pyronin. Feulgen developed a stoichiometric procedure for staining DNA involving a derivatizing dye (fuchsin) to a Schiff base (Feulgen & Rossenback, 1924), which was used for decades in early work identifying cancer cells. It was not unit Coons and Kaplan developed a more accessible method of antibody conjugation in 1950 that fluorescence became a useful technique (Coons & Kaplan, 1950). Fluoresceinated antibodies were few and far between, the first being developed by Coons et al. in 1941 (Coons, Creech, & Jones, 1941). He conjugated β-anthryl isocyanate to antipneumococcus antibodies, which resulted in blue fluorescence under UV light, and further conjugation of fluorescein-4-isocyanate to antipneumococcal antibodies (Coons, Creech, Jones, & Berliner, 1942) which was a clear demonstration that conjugation was possible, but very difficult. Further, at this time, it was necessary to use a carbon arc-lamp for UV microscopy, something that might make our labs a little smoky and rather difficult to use.

Even then the techniques took several days, and significant chemistry to accomplish, a far cry from the 30-min kits that are available to graduate students today! In his own reflections on the development of immunofluorescence, Albert Coons explained in a memoir why fluorescein was selected: "*Fluorescein was chosen as the label because of the brilliance of its fluorescence and because no green-fluorescing materials had been reported in mammalian tissue.*" (Coons, 1961). What an amazing thought process when almost nothing was known about fluorescence at the time. It is totally understandable, then, why fluorescein became the key fluorescent molecule, a legend in flow cytometry, and fortuitously was maximally excited by the 488-nm laser line. It is in this context that fluorescence detection was slated to become one of the most significant tools available for flow cytometry as the technology emerged.

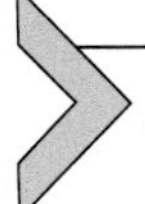

5. Introducing phenotyping for cancer detection by flow cytometry

While Göhde used ethidium bromide for the first published measurement of cellular fluorescence in a flow cytometer (Dittrich & Göhde, 1969a), van Dilla used Feulgen-stained cells (Van Dilla, Trujillo, Mullaney, & Coulter, 1969) in flow in March 1969 and also established the principles of light-scatter analysis by flow cytometry (Mullaney, Van Dilla, Coulter, & Dean, 1969). Several months later Herzenberg used fluorescein diacetate–labeled cells (Hulett et al., 1969) in a flow cytometer, based, as I understand it, on the work of Boris Rotman, who had previously used FDA with a cell culture he obtained from Len Herzenberg (Rotman & Papermaster, 1966). Not long after, Herzenberg combined the cell-sorting concept of Fulwyler (1965) and the fluorescein-antibody conjugation technology of Coons to physically sort antibody-conjugated cells (Bonner, Hulett, Sweet, & Herzenberg, 1972). This was a first from many perspectives; a first use of fluorescence cell sorting, a first combining a laser on a flow cytometer to sort cells, and a first combination of all these to create what was a first fluorescence-based sorting technology and without doubt the initiation of current day technologies. Although the first instrument had only a single detector, they subsequently added scatter-based detectors based on the work of Mullaney et al. (1969). However, there were really two distinct approaches to flow cytometry studies. Clearly Herzenberg was focusing on phenotypic characterization of cell populations, while most others were focused on DNA analysis.

By 1981 there was increased interest in the clinical application of flow cytometry; in a review Laerum noted that one of the first applications was in hematology, particularly in blood and bone marrow (Laerum & Farsund, 1981). However, the approach was the analysis of DNA distribution curves and stains for DNA and RNA. Other potential applications were the analysis of esterases (Parry et al., 1981) and suggestions that light scatter might also be useful in bone-marrow evaluation by flow cytometry as well (Grogan, Scott, & Collins, 1978). The fundamental problem at this time was the limitation by the flow cytometry hardware to one or two colors of fluorescence. Monoclonal antibodies were unavailable commercially until 1979, when Ortho introduced OKT3, OKt4 and OKT8 (Kung, Goldstein, Reinherz, & Schlossman, 1979). Therefore, only nucleic acid dyes, scatter, and a variety of polyclonal conjugated antibodies were available to the clinical flow community. Clearly the focus was on concept such as S-phase monitoring, as this appeared to be relevant to tumor biology (Diamond & Braylan, 1980).

By 1983, there was strong support for flow cytometry as a driving force for cancer diagnosis and some influential authors went as far as stating;

> "Several reports demonstrate increasing morphological immaturity to be associated with increasing DNA content abnormality and increasing S percentage, all of which adversely affect prognosis"
>
> (Barlogie, Raber, et al., 1983).

However, with the advent of monoclonal antibodies the world of flow cytometry moved from DNA to immunology. As monoclonal antibodies became increasingly useful in the flow cytometry scene during the 1980s, multiple companies competed to sell commercial products that enabled a huge growth in flow cytometry. It soon became apparent that few standards were available to ensure that assay design and data interpretation were capable of meeting clinical standards, resulting in quality control implementation (Landay & Muirhead, 1989). To some extent, this was a result of a vast increase in the use of flow cytometry for immunophenpotyping blood cells; standardized staining procedures were clearly needed. By the mid 1990s the concept of evaluating leukemia subtypes appeared to be useful, particularly focused on identifying high proportions of S-phase tumor cells (Orfao et al., 1995). It should be noted that in the mid 1980s, Oi, Glazer, & Stryer (1982) demonstrated that phycobiliprotein conjugates were useful in flow cytometry, creating a pathway for multiparameter analysis, an area that blossomed in more recent times.

By the mid 1990s the concept of evaluating leukemia subtypes appeared to be useful particularly focused on identifying high proportions of S-phase tumor cells (Orfao et al., 1995). However, as monoclonal antibodies arrived on the flow cytometry scene in the 1980s, it became apparent that few standards were available to ensure that assay design and data interpretation were capable of meeting clinical standards resulting in quality control implementation (Landay & Muirhead, 1989). To some extent, this was a result of a vast increase in the use of flow cytometry for immunophenpotyping blood cells and standardized staining procedures were needed. Schlossman had introduced monoclonal antibodies about a decade earlier (Kung et al., 1979) and multiple companies were selling commercial products enabling a huge growth in flow cytometry.

6. Multiparameter flow cytometry for cancer diagnosis

Len Herzenberg's work drove flow cytometry into what is, today, probably the primary focus of flow cytometry: immunophenotyping. Len was fortunate to be doing a sabbatical in the laboratory of César Milstein, where monoclonal antibodies were invented. It was Len who actually suggested the name *hybridoma* for the combination cell that was the basis of the technology, as noted by Milstein (1999) who is worth quoting for our younger colleagues to see both the impact that our flow colleague had and the critical value of sabbaticals for scientists (Köhler and Herzenberg were both doing sabbaticals).

> "The term hybridoma was proposed by Len Herzenberg during a sabbatical in my laboratory in 1976/1977. At a high-table conversation at a Cambridge College, Len was told by one of the dons that hybridoma was garbled Greek. By then however, the term was becoming popular among us, and we decided to stick to it."

Len Herzenberg's own group went on to firmly establish flow cytometry and multiparameter flow cytometry, heavily driven by the instrumentation development by Becton Dickinson Immunocytometry Systems (BDIS) as it was then called. It was not long before Len's group were actively using conjugated antibodies in multiple cell populations. Subsequently, they began a multidecade process of advancing the number of simultaneous fluorescence measurements by flow cytometry. By 1996, Roederer utilized fluorescein, Texas Red, Cy7, APC, as well as conjugated probes such as Cy5-PE and Cy7-PE and Cy7-APC and demonstrated six- color staining of peripheral blood (Roederer, Kantor, Parks, & Herzenberg, 1996). Soon after, the same team showed eight-color staining and interestingly with a 407-nm laser line (although from an Innova 300 Krypton Laser) (Roederer et al., 1997). This advanced to nine colors with cascade blue (Bigos et al., 1999). However, despite the rapid advancement in the number of simultaneous markers for flow cytometry, the clinical side of the business lacked both uniformity of instrumentation and commonality of reagents. For instance, by the end of the 20th century, there were only a very few instruments focused on clinical cytometry, such as the Coulter XL, the BD FACS Calibur and IMAGN 2000, and the CellDyn 4000. The reagent scene was volatile, to say the least. The primary companies were Pharmingen (acquired by BD in 1997) and Coulter Immunology (acquired by Beckman in 1997).

A significant advance in analytical gating was the use of a combination of an antibody stain (CD45) with side scatter. This significantly improved phenotypic determination of blast cells by discriminating between leukemic blast cells and residual normal cells, which could then be excluded from analysis (Lacombe et al., 1997; Stelzer, Shults, & Loken, 1993).

7. The future for clinical cytometry

Two approaches to clinical cytometry are of interest to different groups. One approach is focused on 10-to 12-parameter flow in a traditional instrument. The other suggests the potential for spectral flow cytometry. The former is the present mode, the latter may be the future. It is currently unclear if DNA analysis will re-appear in the clinical domain, but as noted in multiple chapters in this volume, there is little doubt that flow cytometry has a significant role in cancer analysis. However as noted above, in the early days of flow cytometry when instrumentation was capable of only 1–3 parameters, DNA analysis was the most interesting aspect of clinical analysis. The question is, will future technological advances in flow-cytometry technology provide tools for interrogating DNA in a way similar to the tools now available for immunophenotyic analysis?

References

Andreeff, M., Darzynkiewicz, Z., Sharpless, T. K., Clarkson, B. D., & Melamed, M. R. (1980). Discrimination of human leukemia subtypes by flow cytometric analysis of cellular DNA and RNA. *Blood*, *55*, 282–293.

Anonomous. (1957). Cytoanalyzer. *Public Health Reports*, *72*(11), 1038. Retrieved from https://www.ncbi.nlm.nih.gov/pubmed/13485299. https://www.ncbi.nlm.nih.gov/pmc/articles/PMC2031420/pdf/pubhealthreporig00143-0094.pdf.

Avery, O. T., McCarty, M., & MacLeod, M. (1944). Studies on the chemical nature of the substance inducing transformation of pneumococcal types. Induction of transformation by a desoxyribonucleic acid fraction isolated from pneumococcus type III. *The Journal of Experimental Medicine*, *79*, 137–158.

Bagwell, C. B. (2004). DNA histogram analysis for node-negative breast cancer. *Cytometry. Part A*, *58*(1), 76–78. https://doi.org/10.1002/cyto.a.90004.

Bagwell, C. B., Clark, G. M., Spyratos, F., Chassevent, A., Bendahl, P. O., Stal, O., et al. (2001b). Optimizing flow cytometric DNA ploidy and S-phase fraction as independent prognostic markers for node-negative breast cancer specimens. *Cytometry*, *46*(3), 121–135. https://doi.org/10.1002/cyto.1097.

Bagwell, C. B., Clark, G. M., Spyratos, F., Chassevent, A., Bendahl, P. O., Stal, O., et al. (2001a). DNA and cell cycle analysis as prognostic indicators in breast tumors revisited. *Clinics in Laboratory Medicine*, *21*(4), 875–895. Retrieved from https://www.ncbi.nlm.nih.gov/pubmed/11770293.

Bandura, D. R., Baranov, V. I., Ornatsky, O. I., Antonov, A., Kinach, R., Lou, X., et al. (2009). Mass cytometry: Technique for real time single cell multitarget immunoassay based on inductively coupled plasma time-of-flight mass spectrometry. *Analytical Chemistry*, *81*(16), 6813–6822. https://doi.org/10.1021/ac901049w.

Barlogie, B., Maddox, A. M., Johnston, D. A., Raber, M. N., Drewinko, B., Keating, M. J., et al. (1983). Quantitative cytology in leukemia research. *Blood Cells*, *9*(1), 35–55.

Barlogie, B., Raber, M. N., Schumann, J., Johnson, T. S., Drewinko, B., Swartzendruber, D. E., et al. (1983). Flow cytometry in clinical cancer research. *Cancer Research*, *43*, 3982–3997.

Bigos, M., Baumgarth, N., Jager, G. C., Herman, O. C., Nozaki, T., Stovel, R. T., et al. (1999). Nine color eleven parameter immunophenotyping using three laser flow cytometry. *Cytometry*, *36*(1), 36–45.

Bonner, W. A., Hulett, H. R., Sweet, R. G., & Herzenberg, L. A. (1972). Fluorescence activated cell sorting. *Review of Scientific Instruments*, *43*(3), 404–409. https://www.jstor.org/stable/24950310.

Buchner, T., Asseburg, U., Kamanabroo, D., Hiddemann, W., Hiddemann, R. M., Barlogie, B., et al. (1974). Clinical studies on combined chemotherapy of leukemia with partial cell synchronization. *Verhandlungen der Deutschen Gesellschaft für Innere Medizin*, *80*(1674–8), 1674–1678.

Buchner, T., Barlogie, B., Asseburg, U., Hiddemann, W., Kamanabroo, D., & Gohde, W. (1974). Accumulation of S-phase cells in the bone marrow of patients with acute leukemia by cytosine arabinoside. *Blut*, *28*(4), 299–300.

Buchner, T., Gohde, W., Dittrich, W., & Barlogie, B. (1972). Proliferation kinetics of leukemias before and during therapy as based on impulse cytophotometry. *Verhandlungen der Deutschen Gesellschaft für Innere Medizin*, *78*(159–62), 159–162.

Buchner, T., Hiddemann, W., Schneider, R., Kamanabroo, D., & Gohde, W. (1973). Cell-kinetic effects of leukemia therapy in chemical medicine as based on the DNA-histogram with impulse cytophotometry. *Die Medizinische Welt*, *24*(42), 1616–1617.

Casperson, O. (1964). Quantitative cytochemical studies on normal, malignant premalignant and atypical cell populations from the himan uterine cervix. *Acta Cytologica*, *8*, 45.

Caspersson, T. O. (1941). Ribonucleic acids and the synthesis of cellular proteins. *Naturwiss*, *28*, 33.

Caspersson, T. O. (1950). *Cell growth and cell function. A cytochemical study*. New York: W.W. Norton & Company.

Chao, D. L., Sanchez, C. A., Galipeau, P. C., Blount, P. L., Paulson, T. G., Cowan, D. S., et al. (2008). Cell proliferation, cell cycle abnormalities, and cancer outcome in patients with Barrett's esophagus: A long-term prospective study. *Clinical Cancer Research*, *14*(21), 6988–6995 (Retrieved from PM:18980994).

Choi, W. T., Emond, M. J., Rabinovitch, P. S., Ahn, J., Upton, M. P., & Westerhoff, M. (2015). "Indefinite for dysplasia" in Barrett's esophagus: Inflammation and DNA content abnormality are significant predictors of early detection of neoplasia. *Clinical and Translational Gastroenterology*, *6*(3), e81. https://doi.org/10.1038/ctg.2015.7.

Choi, W. T., Rabinovitch, P. S., Wang, D., & Westerhoff, M. (2015). Outcome of "indefinite for dysplasia" in inflammatory bowel disease: Correlation with DNA flow cytometry and other risk factors of colorectal cancer. *Human Pathology*, *46*(7), 939–947. https://doi.org/10.1016/j.humpath.2015.03.009.

Collste, L. G., Darzynkiewicz, Z., Traganos, F., Sharpless, T. K., Sogani, P., Grabstald, H., et al. (1980). Flow cytometry in bladder cancer detection and evaluation using acridine orange metachromatic nucleic acid staining of irrigation cytology specimens. *Journal of Urology*, *123*, 478–485.

Coons, A. H. (1961). The beginnings of immunofluorescence. *The Journal of Immunology*, *87*(5), 499. Retrieved from http://www.jimmunol.org/content/87/5/499.abstract.

Coons, A. H., Creech, H. J., & Jones, R. N. (1941). Immunological properties of an antibody containing a fluorescent group. *Proceedings of the Society for Experimental Biology and Medicine*, *47*, 200–202.

Coons, A. H., Creech, H. J., Jones, R. N., & Berliner, E. (1942). The demonstration of pneumococcal antigen in tissues by the use of fluorescent antibody. *Journal of Immunology*, *45*, 159–170.

Coons, A. H., & Kaplan, M. H. (1950). Localization of antigen in tissue cells. II. Improvements in a method for the detection of antigen by means of fluorescent antibody. *Journal of Experimental Medicine*, *91*, 1–13.

Courtney, W. B., Hilberg, A. W., Ingraham, S. C., 2nd, Kaiser, R. F., Pruitt, J. C., & Bouser, M. M. (1960). Field trial of the cytoanalyzer: 1,184 specimens analyzed. *Journal of the National Cancer Institute*, *24*, 1167–1179. Retrieved from https://www.ncbi.nlm.nih.gov/pubmed/13812396.

Darzynkiewicz, Z., Traganos, F., Sharpless, A. T., & Melamed, M. R. (1976). Lymphocyte stimulation: A rapid multiparameter analysis. *PNAS*, *73*(8), 2881–2884. https://doi.org/10.1073/pnas.73.8.2881.

De Rosa, S. C., Brenchley, J. M., & Roederer, M. (2003). Beyond six colors: A new era in flow cytometry. *Nature Medicine*, *9*(1), 112–117.

Diacumakos, E. G., Day, E., & Kopac, M. J. (1962). Further development of the cytoanalyzer. *Acta Cytologica*, *6*, 238–244. Retrieved from https://www.ncbi.nlm.nih.gov/pubmed/13886085.

Diamond, L. W., & Braylan, R. C. (1980). Flow analysis of DNA content and cell size in non-Hodgkin's lymphoma. *Cancer Research*, *40*(3), 703–712. Retrieved from https://www.ncbi.nlm.nih.gov/pubmed/7471089.

Dittrich, W., & Göhde, W. (1969a). Impulse fluorometry of single cells in suspension. *Zeitschrift für Naturforschung B*, *24*(3), 360–361.

Dittrich, W., & Göhde, W. (1969b). Impulsfluorometrie bei einzelzellen in suspensionen. *Zeitschrift für Naturforschung B*, *24b*, 360–361. https://doi.org/10.1515/znb-1969-0326.

Editorial. (1975). Pulse Cytophotometry. *The Lancet*, *305*(7904), 435–436. https://doi.org/10.1016/S0140-6736(75)91497-X.

Editorial. (1996). Clinical practice guidelines for the use of tumor markers in breast and colorectal cancer. Adopted on May 17, 1996 by the American Society of Clinical Oncology. *Journal of Clinical Oncology, 14*, 2843–2877.

Ehrlich, P., & Lazarus, A. (1900). *Histology of the blood Normal and pathological* (W. Myers Ed.). C. J. Clay and Sons.

Feulgen, R., & Rossenback, H. (1924). Mikroskopisch-chemischer Nachweis einer Nucleinsäure vom Typus der Thymonucleinsäure und die darauf beruhende elektive Färbungvon Zellkernen in mikroskopischen Präparaten. *Hoppe-Seyler's Zeitschrift für Physiologische Chemie, 135*.

Frost, J. K., Tyrer, H. W., Pressman, N. J., Albright, C. D., Vansickel, M. H., & Gill, G. W. (1979). Automatic cell identification and enrichment in lung cancer. I. Light scatter and fluorescence parameters. *Journal of Histochemistry & Cytochemistry, 27*(1), 545–551. https://doi.org/10.1177/27.1.86575.

Fulwyler, M. J. (1965). Electronic separation of biological cells by volume. *Science, 150*, 910–911.

George, T. C., Basiji, D. A., Hall, B. E., Lynch, D. H., Ortyn, W. E., Perry, D. J., et al. (2004). Distinguishing modes of cell death using the ImageStream-Æ multispectral imaging flow cytometer. *Cytometry Part A, 59A*(2), 237–245. Retrieved from https://doi.org/10.1002/cyto.a.20048.

Giesen, C., Wang, H. A., Schapiro, D., Zivanovic, N., Jacobs, A., Hattendorf, B., et al. (2014). Highly multiplexed imaging of tumor tissues with subcellular resolution by mass cytometry. *Nature Methods, 11*(4), 417–422. https://doi.org/10.1038/nmeth.2869.

Gohde, W., Dittrich, W., Zinser, K. H., & Prieshof, J. (1972). Impulse cytophotometric measurements on atypical cell smears from vagina and cervix uteri. *Geburtshilfe und Frauenheilkunde, 32*(5), 382–393.

Grogan, W. M., Scott, R. B., & Collins, J. M. (1978). Light scatter characteristics of erythroid precursor cells studied in flow analysis. *Proceedings of the Society for Experimental Biology and Medicine, 159*(2), 219–222. https://doi.org/10.3181/00379727-159-40318.

Hedley, D. W. (1989). Flow cytometry using paraffin-embedded tissue: Five years on. *Cytometry, 10*, 229–241.

Hedley, D. W., Friedlander, M. L., Taylor, I. W., Rugg, C. A., & Musgrove, E. A. (1983). Method for analysis of cellular DNA content of paraffin-embedded pathological material using flow cytometry. *Journal of Histochemistry & Cytochemistry, 31*(11), 1333–1335. https://doi.org/10.1177/31.11.6619538.

Hulett, H. R., Bonner, W. A., Barrett, J., & Herzenberg, L. A. (1969). Cell sorting: Automated separation of mammalian cells as a function of intracellular fluorescence. *Science, 166*(3906), 747–749. https://doi.org/10.1126/science.166.3906.747.

Ingram, M. L., & Barnes, S. W. (1951). Experimental confirmation of a previoulsly reported unusual finding in the blood of cyclotron workers. *Science, 113*(2924), 32–34.

Ingram, M., & Preston, K., Jr. (1964). Importance of automatic pattern recognition techniques in the early detection of altered blood cell production. *Annals of the New York Academy of Sciences, 113*, 1066–1072. https://doi.org/10.1111/j.1749-6632.1964.tb40724.x.

Ingram, M., & Preston, K., Jr. (1970). Automatic analysis of blood cells. *Scientific American, 223*(5), 72–82. https://doi.org/10.1038/scientificamerican1170-72.

Kamentsky, L. A. (1958). USA patent no. 3106699. USPTO.

Kamentsky, L. A. (1962). USA patent no. 3018471. USPTO.

Kamentsky, L. A., Derman, H., & Melamed, M. R. (1963). Ultraviolet absorption in epidermoid cancer cells. *Science, 142*, 1580–1583.

Kamentsky, L. A., & Melamed, M. R. (1967). Spectrophotometric cell sorter. *Science, 156*, 1364–1365.

Kamentsky, L. A., Melamed, M. R., & Derman, H. (1965). Spectrophotometer: New instrument for ultrarapid cell analysis. *Science*, *150*, 630–631.

Kung, P., Goldstein, G., Reinherz, E. L., & Schlossman, S. F. (1979). Monoclonal antibodies defining distinctive human T cell surface antigens. *Science*, *206*(4416), 347–349. Retrieved from http://www.sciencemag.org/content/206/4416/347.abstract.

Lacombe, F., Durrieu, F., Briais, A., Dumain, P., Belloc, F., Bascans, E., et al. (1997). Flow cytometry CD45 gating for immunophenotyping of acute myeloid leukemia. *Leukemia*, *11*(11), 1878–1886.

Laerum, O. D., & Farsund, T. (1981). Clinical application of flow cytometry: A review. *Cytometry*, *2*(1), 1–13.

Landay, A. L., & Muirhead, K. A. (1989). Procedural guidelines for performing immunophenotyping by flow cytometry. *Clinical Immunology and Immunopathology*, *52*, 48–60.

Melamed, M. R., Adams, L. R., Traganos, F., Zimring, A., & Kamentsky, L. A. (1972). Acridine orange metachromasia for characterization of leukocytes in leukemia, lymphoma, and other neoplasms. *Cancer*, *29*, 1361–1368.

Melamed, M. R., & Kamentsky, L. A. (1969). An assessment of the potential role of automatic devices in cytology screening. *Obstetrical and Gynecological Survey*, *24*, 914–926.

Melamed, M. R., Kamentsky, L. A., & Boyse, E. A. (1969). Cytotoxic test automation: A live-dead cell differential counter. *Science*, *163*, 285–286.

Melamed, M. R., Traganos, F., Sharpless, T., & Darzynkiewicz, Z. (1976). Urinary cytology automation. Preliminary studies with acridine orange stain and flow-through cytofluorometry. *Investigative Urology*, *13*, 331–338.

Mellors, R. C., Keane, J. F., Jr., & Papanicolaou, G. N. (1952). Nucleic acid contents of the squamous cancer cell. *Science*, *116*, 265–269.

Mellors, R. C., & Silver, R. (1951). A microfluorometric scanner for the differential detection of cells: Application to exfoliative cytology. *Science*, *114*(2962), 356–360. Retrieved from C:\Documents\Word\Manuscripts\2011\Portugese lung cancer group\Science-1951-Mellors-356-60.jpg.

Mendelsohn, M. L. (1958). The two-wavelength method of microspectrophotometry. *The Journal of Biophysical and Biochemical Cytology*, *4*(4), 425–431. Retrieved from http://jcb.rupress.org/content/4/4/425.abstract.

Mendelsohn, M., & Mayall, B. H. (1974). *Human chromosome methodology*. New York: Academic Press.

Mendelsohn, M. L., & Richards, B. M. (1958). A comparison of scanning and two-wavelength microspectrophotometry. *The Journal of Biophysical and Biochemical Cytology*, *4*(6), 707–709. Retrieved from http://jcb.rupress.org/content/4/6/707.abstract.

Milstein, C. (1999). The hybridoma revolution: An offshoot of basic research. *BioEssays*, *21*(11), 966–973.

Mullaney, P. F., Van Dilla, M. A., Coulter, J. R., & Dean, P. N. (1969). Cell sizing: A light scattering photometer for rapid volume determination. *Review of Scientific Instruments*, *40*, 1029–1032.

Oi, V. T., Glazer, A. N., & Stryer, L. (1982). Fluorescent phycobiliprotein conjugates for analyses of cells and molecules. *Journal of Cell Biology*, *93*(3), 981–986. https://doi.org/10.1083/jcb.93.3.981.

Orfao, A., Ciudad, J., Gonzalez, M., Lopez, A., del Mar, A. M., Paz Bouza, J. I., et al. (1995). Flow cytometry in the diagnosis of cancer. *Scandinavian Journal of Clinical and Laboratory Investigation. Supplementum.*, *221*(145–52), 145–152.

Papanicolaou, G. N., & Traut, H. F. (1941). The diagnostic value of vaginal smears in carcinoma of the uterus *. *American Journal of Obstetrics & Gynecology*, *42*(2), 193–206. https://doi.org/10.1016/S0002-9378(16)40621-6.

Papanicolaou, G. N., & Traut, H. F. (1943). Diagnosis of uterine cancer by the vaginal smear. *American Journal of the Medical Sciences*, *206*, 811.

Parry, M. F., Root, R. K., Metcalf, J. A., Delaney, K. K., Kaplow, L. S., & Richar, W. J. (1981). Myeloperoxidase deficiency: prevalence and clinical significance. *Annals of Internal Medicine, 95*, 293–301. Retrieved from http://annals.org/article.aspx?articleid=695047.

Reichard, P. (2002). Osvald T. Avery and the Nobel prize in medicine. *Journal of Biological Chemistry, 277*(16), 13355–13362. https://doi.org/10.1074/jbc.R200002200.

Reiffenstuhl, G., Severin, E., Dittrich, W., & Gohde, W. (1971). Impulse cytophotometry of vaginal and cervical smears. *Archiv für Gynäkologie, 211*(4), 595–616.

Robinson, J. P. (2004). *Multispectral cytometry: The next generation* (pp. 36–40). Pittsfield, MA, USA: Biophotonics International, Laurin Publishing.

Robinson, R. D., Wheeless, D. M., Hespelt, S. J., & Wheeless, L. L. (1990). System for acquisition and real-time processing of multidimensional slit-scan flow cytometric data. *Cytometry, 11*(3), 379–385. https://doi.org/10.1002/cyto.990110308.

Roederer, M., De Rosa, S., Gerstein, R., Anderson, M., Bigos, M., Stovel, R., et al. (1997). 8 Color, 10-parameter flow cytometry to elucidate complex leukocyte heterogeneity. *Cytometry, 29*(4), 328–339.

Roederer, M., Kantor, A. B., Parks, D. R., & Herzenberg, L. A. (1996). Cy7PE and Cy7APC: Bright new probes for immunofluorescence. *Cytometry, 24*(3), 191–197. Retrieved from http://onlinelibrary.wiley.com/store/10.1002/(SICI)1097-0320(19960701)24:3%3C191::AID-CYTO1%3E3.0.CO;2-L/asset/1_ftp.pdf?v=1&t=ivyacrpy&s=55e8b824c07a190c1b03322bb22ea74360df6a62.

Rotman, B., & Papermaster, B. W. (1966). Membrane properties of living mammalian cells as studied by enzymatic hydrolysis of fluorogenic esters. *Proceedings of the National Academy of Sciences of the United States of America, 55*(1), 134–141.

Salzman, G. C., Crowell, J. M., Goad, C. A., Hansen, K. M., Hiebert, R. D., LaBauve, P. M., et al. (1975). A flow-system multiangle light-scattering instrument for cell characterization. *Clinical Chemistry, 21*, 1297–1304.

Salzman, G. C., Crowell, J. M., Hansen, K. M., Ingram, M., & Mullaney, P. F. (1976). Gynecologic specimen analysis by multiangle light scattering in a flow system. *Journal of Histochemistry and Cytochemistry, 24*, 308–314.

Salzman, G. C., Crowell, J. M., Martin, J. C., Trujillo, T. T., Romero, A., Mullaney, P. F., et al. (1975). Cell classification by laser light scattering: Identification and separation of unstained leukocytes. *Acta Cytologica, 19*, 374–377.

Salzman, G., Hiebert, R., Jett, J., & Bartholdi, M. (1980). *High-speed single particle sizing by light scattering in a flow system. Vol. 0220*. SPIE.

Schumann, J., Ehring, F., & Göhde, W. (1975). Cell synchronization in solid tumors. *Recent Results in Cancer Research, 52*, 206–214.

Schumann, J., Ehring, F., Gohde, W., & Dittrich, W. (1971). Pulse cytophotometry of DNA in skin tumours. *Archiv für Klinische und Experimentelle Dermatologie, 239*(4), 377–389.

Schumann, J., & Gohde, W. (1974). Cytokinetic effect of bleomycin on mouse Ehrlich carcinoma in vivo. Possibilities of tumor cell synchronization for radiotherapy. *Strahlentherapie, 147*(3), 298–307.

Stelzer, G. T., Shults, K. E., & Loken, M. R. (1993). CD45 gating for routine flow cytometric analysis of human bone marrow specimens. *Annals of the New York Academy of Sciences*, 265–280.

Tolles, W. E. (1955). The cytoanalyzer-an example of physics in medical research. *Transactions of the New York Academy of Sciences, 17*(3), 250–256. https://doi.org/10.1111/j.2164-0947.1955.tb01204.x.

Traganos, F., Darzynkiewicz, Z., Sharpless, T. K., & Melamed, M. R. (1979). Erythroid differentiation of friend leukemia cells as studied by acridine orange staining and flow cytometry. *Journal of Histochemistry and Cytochemistry, 27*, 382–389.

Vaickus, L. J., & Tambouret, R. H. (2015). Young investigator challenge: The accuracy of the nuclear-to-cytoplasmic ratio estimation among trained morphologists. *Cancer Cytopathology*, *123*(9), 524–530. https://doi.org/10.1002/cncy.21585.

Van Dilla, M. A., Trujillo, T. T., Mullaney, P. F., & Coulter, J. R. (1969). Cell microfluorometry: A method for rapid fluorescence measurement. *Science*, *163*, 1213–1214.

Watson, J. D., & Crick, F. H. C. (1953). A structure for deoxyribose nucleic acid. *Nature*, *171*, 737–738.

Wheeless, L. L., Jr., Cambier, J. L., Cambier, M. A., Kay, D. B., Wightman, L. L., & Patten, S. F., Jr. (1979). False alarms in a slit-scan flow system: Causes and occurrence rates. Implications and potential solutions. *The Journal of Histochemistry and Cytochemistry*, *27*(1), 596–599. https://doi.org/10.1177/27.1.374626.

Wheeless, L. L., Jr., Kay, D. B., Cambier, M. A., Cambier, J. L., & Patten, S. F., Jr. (1977). Imaging systems for correlation of false alarms in flow. *The Journal of Histochemistry and Cytochemistry*, *25*(7), 864–869. https://doi.org/10.1177/25.7.330736.

Wheeless, L. L., Jr., & Patten, S. F., Jr. (1973a). Slit-scan cytofluorometry. *Acta Cytologica*, *17*(4), 333–339. Retrieved from https://www.ncbi.nlm.nih.gov/pubmed/4579184.

Wheeless, L. L., Jr., & Patten, S. F., Jr. (1973b). Slit-scan cytofluorometry: Basis for an automated cytopathology prescreening system. *Acta Cytologica*, *17*(5), 391–394. Retrieved from https://www.ncbi.nlm.nih.gov/pubmed/4583034.

Wheeless, L. L., Jr., Reeder, J. E., & O'Connell, M. J. (1990). Slit-scan flow analysis of cytologic specimens from the female genital tract. *Methods in Cell Biology*, *33*, 501–507. https://doi.org/10.1016/s0091-679x(08)60549-x.

Wilson, G. D., & Marples, B. (2007). Flow cytometry in radiation research: Past, present and future. *Radiation Research*, *168*(4), 391–403 (Retrieved from PM:17903043).

CHAPTER TWO

The contribution of automated cytometry in immuno-oncology

Andrea Sbrana[a,b], Giuliano Mazzini[c], Giuditta Comolli[d], Andrea Antonuzzo[e], and Marco Danova[f,g,*]

[a]Department of Surgical, Medical and Molecular Pathology and Critical Care Area, University of Pisa, Pisa, Italy
[b]Service of Pneumo-Oncology, Unit of Pneumology, Pisa, Italy
[c]Institute of Molecular Genetics IGM-CNR, Pavia, Italy
[d]Department of Microbiology and Virology and Laboratory of Biochemistry-Biotechnology and Advanced Diagnostics, IRCCS San Matteo Foundation, Pavia, Italy
[e]Unit of Medical Oncology 1, AOU Pisana, Pisa, Italy
[f]Unit of Internal Medicine and Medical Oncology, Vigevano Civic Hospital, Pavia, Italy
[g]LIUC University, Castellanza, Varese, Italy
*Corresponding author: e-mail address: marco_danova@asst-pavia.it

Contents

Abstract

Cancer immunotherapy has been a real revolution and has given many survival chances to several patients. However, the understanding of resistance to immunotherapy is still an unmet need in clinical practice. Monitoring of immune mechanisms could be a tool to better understand this phenomenon. FCM and CyTOF could be used in this field, since they allow the simultaneous analysis of several protein expressions pattern, thus possibly understanding the functions of several immune cell populations, such as T cells, and their interactions with tumor cells and tumor microenvironment. Furthermore, automated cytometry could be used to understand the interaction of drugs with their target through the analysis of receptor occupancy. Spectral overlap, however, could be a limit for multiple simultaneous analyses. Other possible limitations of these techniques are a low number of cells in samples and the need for viable cells (with the possible

Methods in Cell Biology, Volume 195
ISSN 0091-679X
https://doi.org/10.1016/bs.mcb.2023.03.005

interference of cell debris). The lack of standardized protocols, and thus the difficult reproducibility, have been the major limit to their application in clinical practice, so international efforts have been made to get to shared guidelines. Ongoing trials are to answer to the possibility of clinical application of these techniques.

1. Introduction

Immunotherapy has been a revolution in the field of cancer therapy, being an effective option for many patients (Decker et al., 2017; Farkona, Diamandis, & Blasutig, 2016). However, biomarkers of response to immunotherapy are still lacking (Butterfield, 2017; Yuan et al., 2016).

Translating the principles of cancer immunotherapy into clinical practice is a real challenge. Furthermore, the development of immune monitoring assays is problematic. Automated cytometric techniques (Greenplate, Johnson, Ferrell, & Irish, 2016; Maecker & Harari, 2015) might represent an interesting option in this field.

In the 1960s, the analytic technique of flow cytometry (FCM) was introduced to measure various characteristics of single cells in suspension after the excitation with a light source (Blow, 2017; Robinson & Roederer, 2015). The study of DNA content to analyze cell ploidy or cell proliferation rate was the main focus of the first published trials in the field of medical oncology (Mazzini & Danova, 2017). The development of monoclonal antibodies (moAbs) and the introduction of new fluorescent dyes with narrow excitation and emission spectra allowed for a wide application of this technology, especially in the field of blood neoplasms (Adan, Alizada, Kiraz, Baran, & Nalbant, 2017; Craig & Foon, 2008). Recently, FCM was used in the study of the so-called "rare events," such as the presence of residual leukemic blasts in the bone marrow after treatment, blood dendritic cells or those cell types associated with a metastatic event, such as endothelial progenitor cells and circulating tumor cells, has become more and more promising (Irish & Doxie, 2014; Liang & Fu, 2017; Proserpio & Lönnberg, 2016).

A recent new analytical approach is mass cytometry. It combines the precision of mass spectrometry with the power of flow cytometric analysis. This technique is expected to answer a multitude of biological questions with a single sample so that complex cell systems may be easier to study to develop possible biomarkers. However, at the moment the quantity and complexity of data obtained with this technology require some analytical considerations (Behbehani, 2017; Spitzer & Nolan, 2016).

We try to review some of the current clinical applications of FCM in Oncology and discuss the potential of both FCM and mass cytometry in the field of immuno-oncology.

2. Cancer immunotherapy today

Immunotherapy is a treatment that aims at activating the patient's immune system to fight against cancer cells. Several attempts in the past brought unsatisfactory results, but recently the introduction of MoAbs has brought to a new golden era for cancer immunotherapy (Chen, Bode, & Dong, 2017). Its success mainly depends on a complex interaction between host immune cells and the tumor microenvironment. Because of the complexity of this phenomenon, it is important to better know them to maximize the efficacy of immunotherapy.

Immunotherapy may be based on the stimulation of effector mechanisms or the inhibition of immunosuppressive events. The first strategy might be obtained through the vaccination with cancer antigens or the enhancement of antigen-presenting mechanisms (Ozverel, Karaboz, & Nalbantsoy, 2017), but it also includes adoptive cellular therapy (e.g., administering immunocompetent cells directly to patients), the administration of oncolytic viruses, and the use of Abs that enhance T cell activity (Eggermont, Paulis, Tel, & Figdor, 2014). However, the inhibition strategy is the one that is mostly used nowadays and it includes the use of MoAbs against immune checkpoint inhibitors, such as cytotoxic T-lymphocyte antigen 4 (CTLA-4) and programmed cell death protein 1 (PD-1)/PD-ligand 1 (PD-L1) (Śledzińska, Menger, Bergerhoff, Peggs, & Quezada, 2015), that are currently approved for several indications and are associated with clinically-significant benefit and manageable tolerability profile.

Ipilimumab is a fully human immunoglobulin G1 (IgG1) MoAb against CTLA-4. It demonstrated a survival benefit in metastatic melanoma (Hodi et al., 2010) with a consistent proportion of treated patients alive after 10 years (Schadendorf et al., 2013).

The first anti-PD-1 MoAb was Nivolumab, whose efficacy has been proven firstly in melanoma (Robert et al., 2015), but also in other neoplasms, such as non-small cell lung cancer (Brahmer et al., 2015), head and neck squamous cell carcinoma (Ferris et al., 2016), and Hodgkin's lymphoma (Armand et al., 2018). Similarly, Pembrolizumab, another anti-PD-1 antibody, was proven to be effective in several neoplasms, such as untreated metastatic non-small cell lung cancer with a high PD-L1 expression (≥50%) (Reck et al., 2016).

PD-1/PD-L1 axis can also be inhibited with anti-PD-L1 inhibitors, such as Atezolizumab, which was proven to be effective in previously treated NSCLC patients (Fehrenbacher et al., 2016), or Avelumab, which is approved for the use in advanced Merkel-cell carcinoma (Kaufman et al., 2016).

In the last years, cancer immunotherapy has also progressed with the beginning of clinical trials that tried to understand the role of immunotherapy in combination with other therapeutic options, such as targeted therapies, chemotherapy, and radiotherapy. Nowadays combinations of immunotherapy with targeted therapies are approved for use in first-line metastatic renal-cell carcinoma, where axitinib is used in combination with avelumab (Motzer et al., 2019) or pembrolizumab (Rini et al., 2019). Different immune checkpoint inhibitors are also combined in several diseases, such as melanoma (Wolchok et al., 2017) and renal cell carcinoma (Motzer et al., 2018).

Despite the availability of several MoAbs, the development of predictive tools useful in patient selection has not been equally successful (Gridelli et al., 2017; Liu, Wang, & Bindeman, 2017). These tools might prevent the treatment of unresponsive patients, which are around 40–60% of all treated patients (Maleki Vareki, Garrigós, & Duran, 2017): this would mean avoiding useless toxicity problems, but also the use of expensive therapies without clinical benefit.

At the moment there is no validated pre-treatment biomarker that can guide decision-making. Some insights have emerged from the identification of some post-treatment immune responses that are likely to correlate with clinical outcomes but are of marginal usefulness in clinical practice (Quandt et al., 2017; Roussel et al., 2017).

The difficulty of finding validated biomarkers is also caused by the fact that the immune system is dynamic and its regulation is given by a complex interaction of several elements (Hegde, Karanikas, & Evers, 2016; Martens et al., 2016; Masucci et al., 2016).

On the other hand, the development of new technologies might support and facilitate the finding of biomarkers and their translation into clinical practice (Krieg et al., 2018). These new technologies might try to answer some needs that are a possible obstacle in this field, such as (i) rapid characterization of the tumor and its immune microenvironment at the time of diagnosis; (ii) prediction models of therapy outcome and (iii) characterization of new tools for treatment.

3. Flow cytometry in clinical oncology

3.1 DNA ploidy and cell cycle analysis

The analysis of nuclear DNA content for the evaluation of both ploidy and cell cycle profile has been applied to several malignancies and this analysis have been made through FCM for more than 25 years (Adan et al., 2017; Danielsen, Pradhan, & Novelli, 2016; Mazzini & Danova, 2017). The beginning of such an approach was in the field of blood malignancies, but the development of reliable tissue disaggregation and staining techniques, together with increasingly sophisticated multiparametric analyses supported by advanced acquisition systems and dedicated software for data elaboration and display, has allowed its application to most solid cancers. A wide variety of published experiences describes its potential clinical impact; in particular, the possible prognostic role of tumor ploidy and proliferative activity has been underlined by these trials. The first published investigations focused on the possible correlation of these parameters with known prognostic factors, such as clinical stage or histological grade of differentiation. Later on, tumor ploidy or proliferative activity have been directly correlated with the patients' clinical outcomes. These trials showed that these parameters have major implications in most human malignancies, even when there is no direct correlation with classical clinical or pathological parameters or with prognosis (Mazzini & Danova, 2017). The limitation of FCM use in clinical practice is also related to its level of sensitivity when used to determine the so-called near-diploid/aneuploidy (Mishra et al., 2017). Fewer published experiences are available for the study of tumor proliferative activity (S-phase fraction). Furthermore, flow cytometry studies have also focused on the possibility to predict the efficacy of some treatments and then on the possibility of clinical selection (Malcovati et al., 2013; Pinto, Pereira, Silva, & André, 2017). A major limitation of all these overmentioned studies is that these analyses should be performed in series of homogenous patients within ad hoc clinical trials, whereas the majority of these trials are based on heterogeneous series of patients within not-well-designed clinical experiences. Furthermore, most of these trials were based on formalin-fixed, paraffin-embedded tumor specimens. Three major limitations are related to the use of this kind of material. First, Fixation induces variations in fluorochrome binding and no external diploid standards can be used for the comparison. Second, enzymatic digestion required to produce single-cell suspension causes the presence of subcellular debris, which may overestimate tumor S-phase fraction. Last, the resolution of DNA histograms is lower when compared with fresh materials and then the risk of misclassification of aneuploidy tumors with near-diploid DNA content is higher. All these things considered, DNA content analysis with FCM has no real impact on clinical practice.

3.2 Immunophenotypic analysis

FCM is routinely used for the identification and quantitation of cellular antigens (Craig & Foon, 2008). FCM can evaluate the simultaneous expression of several antigens, as well as the physical properties (size and cytoplasmic complexity) of individual cells, and identify both normal and abnormal cell populations. From the discovery of T- and B-cells, this approach has expanded to the analysis of several other cell types, such as the monocyte-macrophage system or myeloid cells. FCM became a standard in the characterization of immature precursors in myeloid-line neoplasms (with the inclusion of myelodysplastic syndromes) and the classification of lymphatic cell malignancies. In the last years, we have observed impressive progress in the availability of antibodies and fluorochromes, which expanded FCM application. Its application has also moved from diagnosis to therapy monitoring and post-bone marrow transplantation monitoring. It is then an indispensable tool for the diagnosis, classification, staging, and therapy monitoring in onco-hematology (Flores-Montero et al., 2017; Tognarelli, Jacobs, Staiger, & Ullrich, 2016).

3.3 "Rare event" analysis

The analysis of heterogeneous cell populations is quite common. Within these populations "rare" cells can be found. A rare cell is so classified when it represents <0.001% of the entire analyzed population. Its analysis is intuitively complex since artefacts are common when large numbers of cells are acquired and need analyzing. FCM has allowed rare events to be analyzed. For example, the study of minimal residual disease (MRD) in blood malignancies, the study of blood-circulating dendritic cell subsets or circulating endothelial cells have been implemented in clinical practice. In the very last years, this approach has also allowed the study of circulating tumor cells (Irish & Doxie, 2014; Liang & Fu, 2017; Proserpio & Lönnberg, 2016).

3.4 Minimal residual disease in onco-hematology

The presence of MRD after induction therapy in blood malignancies is an important prognostic element, predicting relapse and survival rates, and FCM has allowed its study. A large variety of antibodies and new fluorochromes is available in this field and this allows a better definition of abnormal populations. Furthermore, the improvements in FCM hardware has brought the possibility of collecting high amounts of data. In particular, both the possibility to use new fluorochromes and the low background signal have represented the key for the detection of residual tumor populations. In several malignancies, all these technological improvements have allowed identifying populations with a frequency of $\leq 0.001\%$ (Luskin & Stone, 2017). The use of FCM is then implemented in clinical practice and several clinical protocols are FCM-based: FCM results are then used to change therapy (Manzoni et al., 2010).

3.5 Circulating dendritic cells

Dendritic cells (DC) are specialized phagocytes that activate adaptive immune responses and regulate immunological tolerance according to variations in their micro-environment. They are crucial antigen-presenting cells, then their role in the antitumor immune response is fundamental: they activate T cell-dependent immune response. Peripheral blood DCs are classified into two subsets: (i) CD11c-negative, CD123-positive, lymphoid-derived DCs, and (ii) CD11c-positive, CD123-negative, myeloid-derived DCs. In the last years, they have been studied through FCM to understand their role in different human malignancies (Manzoni et al., 2012), in particular on their role in the immune response of patients with advanced cancer. This might help in the selection of patients that might be a candidate to active immunotherapy protocols (Flores-Montero et al., 2017).

3.6 Circulating endothelial cells and endothelial progenitors

Circulating endothelial cells (CECs) and endothelial progenitor cells (EPCs) are blood-circulating cell populations involved in the process of vasculogenesis and angiogenesis (Danova, Comolli, Manzoni, Torchio, & Mazzini, 2016), so they have been studied as possible biomarkers of cancer neo-angiogenesis. They are known to affect the disease status and the response to anti-cancer treatments (Danova, Torchio, & Mazzini, 2011; De Biasi et al., 2018). Their study through multiparameter FCM is common, but no standardized protocol is known and widely accepted. The presence of multiple protocols, the heterogeneity of the processing phases and the lack of a standardized panel of monoclonal antibodies for their study are the major issues and the major obstacle for their application in clinical practice (De Biasi et al., 2018).

3.7 Circulating tumor cells

The innovation of circulating tumor cells (CTCs) is the fact that they provide a very little invasive source of tumor materials that can be accessed throughout the different phases of the management of the disease, providing a possible biomarker of response to therapy and resistance to treatments (Chen et al., 2017; Kowalik, Kowalewska, & Góźdź, 2017; Wu, Wu, Zhao, Liu, & Li, 2018). CTCs are studied in terms of phenotypic analysis, but also of DNA genotyping and transcriptome evaluation (D'Errico, Machado, & Sainz, 2017). It is known that patients with a higher number of CTCs are associated with a worse prognosis and this has been proven in breast, prostate, and colorectal cancer. The mere evaluation of their number, however, might be not sufficient for a thorough analysis. CTCs are very heterogeneous and can experience wide changes, especially as a response to treatment pressure. CTC characterization is then fundamental and may provide precious information about tumor behavior. Their study can be faced with several technologies (Wu et al., 2018). FCM can be a valid technique for their study (Kowalik et al., 2017). In recent years, different automated cytometry techniques have been employed for CTC characterization. However, even if CTCs are already used in clinical trials as a possible biomarker, their routine clinical utility is still debated.

4. Conclusions

Immune mechanisms monitoring can be a valuable tool to better the success of cancer immunotherapy and to understand both efficacy and resistance mechanisms. Immunotherapy has rapidly progressed, but this has highlighted the real need for technologies that might be used for rare event identification, analysis of change in cell signaling and antigen evaluation. These events require the use of more sensitive and specific technologies, but also of time-sparing techniques.

FCM is one of the preferred technologies in this field since it can analyze protein expression patterns, but also some functions of the cell populations.

FCM has already been used in pre-clinical trials, as well as clinical trials on patients. Its application can help to understand the processes of tumor antigen interactions with T cell systems, the effects of cytokines, the proliferation and assessment of immune cells and the processes of cytotoxicity (Shibru et al., 2021).

Immunotherapy is based on the alteration of immune responses: monoclonal antibodies, by binding to receptors on the surface of several immune cells, either potentiate or inhibit their functions, thus leading to a final potentiation of the immune activity against tumor cells.

Since the beginning of the first immunotherapy trials, automated cytometry has been explored as a possible tool to better immunotherapy management in clinical practice. Cytometry assays might help to identify possible biomarkers to study host immune responses and then to determine therapy efficacy and subsequent treatment change choices. Cytometry could help in the study of tumor micro-environment, whose role in the determination of the efficacy of immunotherapy has been widely accepted (D'Errico et al., 2017).

A new platform has been developed and it is based on the use of mass spectrometry. The combination of flow cytometry and mass spectrometry, called mass cytometry (or cytometry by time-of-flight, CyTOF), allows the simultaneous evaluation of over 40 parameters, thus allowing the study of complex systems, such as the tumor cell and the immune microenvironment. CyTOF is based on the replacement of fluorescent probes with heavy metal isotopes, chelated to a polymer, covalently linked to an antibody. After staining by these probes, the cell suspension is introduced via an aerosol stream onto a plasma column, resulting in ionization of the cells. The heavy metal ions are then focused on a time-to-flight detector (Spitzer & Nolan, 2016; Tanner, Baranov, Ornatsky, Bandura, & George, 2013).

The process of development of new target therapies needs to take into consideration the capability of the drug to engage with its target. FCM could be an appropriate tool to study this phenomenon through the use of receptor occupancy assays (ROA). FCM could allow the identification of specific cell subsets and non-abundant cell antigens.

Receptor occupancy, the ratio between the amount of bound drug and the amount of total receptor on single cells, is a biomarker for treatment response to therapeutic monoclonal antibodies. Receptor occupancy is traditionally measured by flow cytometry. ROA could allow evaluating both receptor expression and drug occupancy, thus leading to fundamental information on efficacy and safety (Audia, Bannish, Bunting, & Riveley, 2021). These things considered, however, spectral overlap can be the limit for multiple simultaneous analyses. This restricts receptor occupancy assays to the analysis of major cell types, although rare cell populations are of potential therapeutic relevance

The measurement of multiple markers in FCM could allow for receptor occupancy evaluation, but also the correct phenotyping of all blood cells, which could allow for a better understanding of drug activity and efficacy.

However, variation of the sensitivity of mass cytometry could lead to an incorrect analysis of receptor occupancy, especially when the receptor and the drug are analyzed through different channels. Recently, the use of antibody-binding quantum simply cellular (QSC) beads has been described as a possible way to optimize the use of mass cytometry for receptor occupancy analysis (Bringeland et al., 2019).

Some technical and methodological issues are to be underlined. A loss of cells in sample preparation is possible both with both FCM and CyTOF, even if it is more probable with CyTOF, which requires more washing steps in the preparation phase. For this reason, samples with $<10^5$ cells are to be considered not appropriate. CyTOF has a lower collection speed (300–500 events/second vs several thousand events/second). Cell debris might interfere with data interpretation. Finally, viable cells are needed, so that overnight blood shipping might compromise the result of the analysis.

The application of FCM to the study of immunotherapy biomarkers and, more generally, of immunological phenomena, as previously mentioned, have suffered from the lack of standardized protocol that could make these assays reproducible and applicable in clinical practice. In the last years, several researchers have tried to draw some guidelines for the application of flow cytometry and cell sorting in immunological studies (Cossarizza, Chang, Radbruch, Acs, et al., 2019). This international effort, made by international experts in the field, is a comprehensive evaluation of all the aspects concerning the study of all major immune cells through the application of the latest cytometry techniques.

Mass cytometry is a novel technique of great interest. Combining mass spectrometry and FCM, it might be interesting for the capability of simultaneous protein expression analysis, even in single cells. CyTOF is a novel technology based on the use of heavy metal ion-labelled probes that are detected by a time-of-flight mass spectrometry and has the capability of detecting up to 40 simultaneous parameters.

Both techniques, FCM and CyTOF, are very promising and several experiences are ongoing to understand their application to clinical practice. Their application, however, needs to be consistent and reproducible, so particular attention to trial design, reagent selection and instrumentation is required.

However, to make FCM and CyTOF applications successful, lots of different factors are involved. The main issues are about the pre-analytical phase with sample preparation, the titrations of the used antibodies, and the application of accurate controls. Furthermore, the high quality of instrument voltages, lasers, and run is fundamental to make these techniques applicable and reproducible (Laskowski, Hazen, Collazo, & Haviland, 2020).

Hence, the major limitation to their applications is represented by their reproducibility, which is not completely satisfying because of the lack of standardized protocols for sample collection, instrument setting, and analysis strategies. Once these drawbacks will be overcome, the two approaches may offer important help to the study of immunotherapy biomarkers to lead to better patient selection and therapy choices.

References

Adan, A., Alizada, G., Kiraz, Y., Baran, Y., & Nalbant, A. (2017). Flow cytometry: Basic principles and applications. *Critical Reviews in Biotechnology*, *37*, 163–176. https://doi.org/10.3109/07388551.2015.1128876.

Armand, P., Engert, A., Younes, A., Fanale, M., Santoro, A., Zinzani, P. L., et al. (2018). Nivolumab for relapsed/refractory classic Hodgkin lymphoma after failure of autologous hematopoietic cell transplantation: Extended follow-up of the multicohort single-arm phase II CheckMate 205 Trial. *Journal of Clinical Oncology*, *36*(14), 1428–1439. https://doi.org/10.1200/JCO.2017.76.0793. Epub 2018 Mar 27. Erratum in: Journal of Clinical Oncology 2018 Sep 10;36(26):2748. PMID: 29584546; PMCID: PMC6075855.

Audia, A., Bannish, G., Bunting, R., & Riveley, C. (2021). Flow cytometry and receptor occupancy in immune-oncology. *Expert Opinion on Biological Therapy*, *22*, 1–8. https://doi.org/10.1080/14712598.2021.1944098. Epub ahead of print. PMID: 34139906.

Behbehani, G. K. (2017). Applications of mass cytometry in clinical medicine: The promises and perils of clinical CyTOF. *Clinics in Laboratory Medicine*, *37*, 945–964. https://doi.org/10.1016/j.cll.2017.07.010.

Blow, N. (2017). Going with the flow. *BioTechniques*, *62*, 201–205. https://doi.org/10.2144/000114543.

Brahmer, J., Reckamp, K. L., Baas, P., Crinò, L., Eberhardt, W. E., Poddubskaya, E., et al. (2015). Nivolumab versus docetaxel in advanced squamous cell-non-small cell lung cancer. *The New England Journal of Medicine*, *373*, 123–135. https://doi.org/10.1056/NEJMoa1504627.

Bringeland, G. H., Bader, L., Blaser, N., Budzinski, L., Schulz, A. R., Mei, H. E., et al. (2019). Optimization of receptor occupancy assays in mass cytometry: Standardization across channels with QSC beads. *Cytometry A*, *95*, 314–322. https://doi.org/10.1002/cyto.a.23723. Epub 2019 Jan 27. PMID: 30688025; PMCID: PMC6590231.

Butterfield, L. H. (2017). The society for immunotherapy of cancer biomarkers task force recommendations review. *Seminars in Cancer Biology* (Epub ahead of print).

Chen, L., Bode, A. M., & Dong, Z. (2017). Circulating tumor cells: Moving biological insights into detection. *Theranostics*, 7, 2606–2619. https://doi.org/10.7150/thno.18588.

Cossarizza, A., Chang, H. D., Radbruch, A., Acs, A., et al. (2019). Guidelines for the use of flow cytometry and cell sorting in immunological studies (second edition). *European Journal of Immunology*, *49*, 1457–1973. https://doi.org/10.1002/eji.201970107. PMID: 31633216; PMCID: PMC7350392.

Craig, F. E., & Foon, K. A. (2008). Flow cytometric immunophenotyping for hematologic neoplasms. *Blood*, *111*, 3941–3967. https://doi.org/10.1182/blood-2007-11-120535.

Danielsen, H. E., Pradhan, M., & Novelli, M. (2016). Revisiting tumor aneuploidy-the place of ploidy assessment in the molecular era. *Nature Reviews. Clinical Oncology*, *13*, 291–304. https://doi.org/10.1038/nrclinonc.2015.208.

Danova, M., Comolli, G., Manzoni, M., Torchio, M., & Mazzini, G. (2016). Flow cytometric analysis of circulating endothelial cells and endothelial progenitors for clinical purposes in oncology: A critical evaluation. *Molecular and Clinical Oncology*, *4*, 909–917. https://doi.org/10.3892/mco.2016.823.

Danova, M., Torchio, M., & Mazzini, G. (2011). Isolation of rare circulating tumor cells in cancer patients: Technical aspects and clinical implications. *Expert Review of Molecular Diagnostics*, *11*, 473–485. https://doi.org/10.1586/erm.11.33.

De Biasi, S., Gibellini, L., Feletti, A., Pavesi, G., Bianchini, E., Lo Tartaro, D., et al. (2018). High speed flow cytometry allows the detection of circulating endothelial cells in hemangioblastoma patients. *Methods*, *134–135*, 1–10.

Decker, W. K., da lva, R. F., Sanabria, M. H., Angelo, L. S., Guimarães, F., Burt, B. M., et al. (2017). Cancer immunotherapy: Historical perspective of a clinical revolution and emerging preclinical animal models. *Frontiers in Immunology*, *8*, 829. https://doi.org/10.3389/fimmu.2017.00829.

D'Errico, G., Machado, H. L., & Sainz, B., Jr. (2017). A current perspective on cancer immune therapy: Step-by-step approach to constructing the magic bullet. *Clinical and Translational Medicine*, *6*, 3. https://doi.org/10.1186/s40169-016-0130-5.

Eggermont, L. J., Paulis, L. E., Tel, J., & Figdor, C. G. (2014). Towards efficient cancer immunotherapy: Advances in developing artificial antigen-presenting cells. *Trends in Biotechnology*, *32*, 456–465. https://doi.org/10.1016/j.tibtech.2014.06.007.

Farkona, S., Diamandis, E. P., & Blasutig, I. M. (2016). Cancer immunotherapy: The beginning of the end of cancer? *BMC Medicine*, *14*, 73. https://doi.org/10.1186/s12916-016-0623-5.

Fehrenbacher, L., Spira, A., Ballinger, M., Kowanetz, M., Vansteenkiste, J., Mazieres, J., et al. (2016). Atezolizumab versus docetaxel for patients with previously treated non-small-cell lung cancer (POPLAR): A multicentre, open-label, phase 2 randomised controlled trial. *Lancet*, *387*, 1837–1846. https://doi.org/10.1016/S0140-6736(16)00587-0.

Ferris, R. L., Blumenschein, G., Jr., Fayette, J., Guigay, J., Colevas, A. D., Licitra, L., et al. (2016). Nivolumab for recurrent squamous-cell carcinoma of the head and neck. *The New England Journal of Medicine*, *375*(19), 1856–1867. https://doi.org/10.1056/NEJMoa1602252. Epub 2016 Oct 8. PMID: 27718784; PMCID: PMC5564292.

Flores-Montero, J., Sanoja-Flores, L., Paiva, B., Puig, N., García-Sánchez, O., Böttcher, S., et al. (2017). Next generation flow for highly sensitive and standardized detection of minimal residual disease in multiple myeloma. *Leukemia*, *31*, 2094–2103. https://doi.org/10.1038/leu.2017.29.

Greenplate, A. R., Johnson, D. B., Ferrell, P. B., Jr., & Irish, J. M. (2016). Systems immune monitoring in cancer therapy. *European Journal of Cancer*, *61*, 77–84. https://doi.org/10.1016/j.ejca.2016.03.085.

Gridelli, C., Ardizzoni, A., Barberis, M., Cappuzzo, F., Casaluce, F., Danesi, R., et al. (2017). Predictive biomarkers of immunotherapy for non-small cell lung cancer: Results from an experts panel meeting of the italian association of thoracic oncology. *Translational Lung Cancer Research*, *6*, 373–386. https://doi.org/10.21037/tlcr.2017.05.09.

Hegde, P. S., Karanikas, V., & Evers, S. (2016). The where, the when, and the how of immune monitoring for cancer immunotherapies in the era of checkpoint inhibition. *Clinical Cancer Research*, *22*, 1865–1874. https://doi.org/10.1158/1078-0432.CCR-15-1507.

Hodi, F. S., O'Day, S. J., McDermott, D. F., Weber, R. W., Sosman, J. A., Haanen, J. B., et al. (2010). Improved survival with ipilimumab in patients with metastatic melanoma. *The New England Journal of Medicine*, *363*, 711–723. https://doi.org/10.1056/NEJMoa1003466.

Irish, J. M., & Doxie, D. B. (2014). High-dimensional single-cell cancer biology. *Current Topics in Microbiology and Immunology*, *377*, 1–21.

Kaufman, H. L., Russell, J., Hamid, O., Bhatia, S., Terheyden, P., D'Angelo, S. P., et al. (2016). Avelumab in patients with chemotherapy-refractory metastatic Merkel cell carcinoma: A multicentre, single-group, open-label, phase 2 trial. *The Lancet. Oncology*, *17*(10), 1374–1385. https://doi.org/10.1016/S1470-2045(16)30364-3. Epub 2016 Sep 1. PMID: 27592805; PMCID: PMC5587154.

Kowalik, A., Kowalewska, M., & Góźdź, S. (2017). Current approaches for avoiding the limitations of circulating tumor cells detection methods-implications for diagnosis and treatment of patients with solid tumors. *Translational Research*, *185*, 58–84.e15. https://doi.org/10.1016/j.trsl.2017.04.002.

Krieg, C., Nowicka, M., Guglietta, S., Schindler, S., Hartmann, F. J., Weber, L. M., et al. (2018). High-dimensional single-cell analysis predicts response to anti-PD-1 immunotherapy. *Nature Medicine*, *24*, 1–153. https://doi.org/10.1038/nm.4466.

Laskowski, T. J., Hazen, A. L., Collazo, R. S., & Haviland, D. (2020). Rigor and reproducibility of cytometry practices for immuno-oncology: A multifaceted challenge. *Cytometry. Part A*, *97*(2), 116–125. https://doi.org/10.1002/cyto.a.23882. Epub 2019 Aug 27. PMID: 31454153.

Liang, S. B., & Fu, L. W. (2017). Application of single-cell technology in cancer research. *Biotechnology Advances*, *35*, 443–449. https://doi.org/10.1016/j.biotechadv.2017.04.001.

Liu, D., Wang, S., & Bindeman, W. (2017). Clinical applications of PD-L1 bioassays for cancer immunotherapy. *Journal of Hematology & Oncology*, *10*, 110. https://doi.org/10.1186/s13045-017-0541-9.

Luskin, M. R., & Stone, R. M. (2017). Can minimal residual disease determination in acute myeloid leukemia be used in clinical practice? *Journal of Oncology Practice/American Society of Clinical Oncology*, *13*, 471–480. https://doi.org/10.1200/JOP.2017.021675.

Maecker, H. T., & Harari, A. (2015). Immune monitoring technology primer: Flow and mass cytometry. *Journal for Immunotherapy of Cancer*, *3*, 44. https://doi.org/10.1186/s40425-015-0085-x.

Malcovati, L., Hellström-Lindberg, E., Bowen, D., Adès, L., Cermak, J., Del Cañizo, C., et al. (2013). Diagnosis and treatment of primary myelodysplastic syndromes in adults: Recommendations from the European LeukemiaNet. *Blood*, *122*, 2943–2964. https://doi.org/10.1182/blood-2013-03-492884.

Maleki Vareki, S., Garrigós, C., & Duran, I. (2017). Biomarkers of response to PD-1/PD-L1 inhibition. *Critical Reviews in Oncology/Hematology*, *116*, 116–124. https://doi.org/10.1016/j.critrevonc.2017.06.001.

Manzoni, M., Mariucci, S., Delfanti, S., Rovati, B., Ronzoni, M., Loupakis, F., et al. (2012). Circulating endothelial cells and their apoptotic fraction are mutually independent predictive biomarkers in Bevacizumab-based treatment for advanced colorectal cancer. *Journal of Cancer Research and Clinical Oncology*, *138*, 1187–1196. https://doi.org/10.1007/s00432-012-1190-6.

Manzoni, M., Rovati, B., Ronzoni, M., Loupakis, F., Mariucci, S., Ricci, V., et al. (2010). Immunological effects of bevacizumab-based treatment in metastatic colorectal cancer. *Oncology*, *79*, 187–196. https://doi.org/10.1159/000320609.

Martens, A., Wistuba-Hamprecht, K., Geukes Foppen, M., Yuan, J., Postow, M. A., Wong, P., et al. (2016). Baseline peripheral blood biomarkers associated with clinical outcome of advanced melanoma patients treated with Ipilimumab. *Clinical Cancer Research*, *22*, 2908–2918. https://doi.org/10.1158/1078-0432.CCR-15-2412.

Masucci, G. V., Cesano, A., Hawtin, R., Janetzki, S., Zhang, J., Kirsch, I., et al. (2016). Validation of biomarkers to predict response to immunotherapy in cancer: Volume I-pre-analytical and analytical validation. *Journal for Immunotherapy of Cancer*, *4*, 76. https://doi.org/10.1186/s40425-016-0178-1.

Mazzini, G., & Danova, M. (2017). Fluorochromes for DNA staining and quantitation. *Methods in Molecular Biology, 1560*, 239–259. https://doi.org/10.1007/978-1-4939-6788-9_18. Springer Science Business Media LLC.

Mishra, S., Awasthi, N. P., Husain, N., Anand, A., Pradeep, Y., Ansari, R., et al. (2017). Flow cytometric analysis of DNA ploidy in liquid based cytology for cervical pre-cancer and cancer. *Asian Pacific Journal of Cancer Prevention, 18*, 1595–1601.

Motzer, R. J., Penkov, K., Haanen, J., Rini, B., Albiges, L., Campbell, M. T., et al. (2019). Avelumab plus axitinib versus sunitinib for advanced renal-cell carcinoma. *The New England Journal of Medicine, 380*(12), 1103–1115. https://doi.org/10.1056/NEJMoa1816047. Epub 2019 Feb 16. PMID: 30779531; PMCID: PMC6716603.

Motzer, R. J., Tannir, N. M., DF, M. D., Arén Frontera, O., Melichar, B., Choueiri, T. K., et al. (2018). Nivolumab plus Ipilimumab versus Sunitinib in Advanced Renal-Cell Carcinoma. *New England Journal of Medicine, 378*(14), 1277–1290. https://doi.org/10.1056/NEJMoa1712126. Epub 2018 Mar 21. PMID: 29562145; PMCID: PMC5972549.

Ozverel, C. S., Karaboz, I., & Nalbantsoy, A. (2017). Novel treatment strategies in cancer immunotherapy. *Acta Biologica Turcica, 30*, 36–51.

Pinto, A. E., Pereira, T., Silva, G. L., & André, S. (2017). Prognostic relevance of DNA flow cytometry in breast cancer revisited: The 25-year experience of the portuguese institute of oncology of lisbon. *Oncology Letters, 13*, 2027–2033. https://doi.org/10.3892/ol.2017.5718.

Proserpio, V., & Lönnberg, T. (2016). Single-cell technologies are revolutionizing the approach to rare cells. *Immunology and Cell Biology, 94*, 225–229. https://doi.org/10.1038/icb.2015.106.

Quandt, D., Dieter Zucht, H., Amann, A., Wulf-Goldenberg, A., Borrebaeck, C., Cannarile, M., et al. (2017). Implementing liquid biopsies into clinical decision making for cancer immunotherapy. *Oncotarget, 8*, 48507–48520. https://doi.org/10.18632/oncotarget.17397.

Reck, M., Rodríguez-Abreu, D., Robinson, A. G., Hui, R., Csőszi, T., Fülöp, A., et al. (2016). Pembrolizumab versus chemotherapy for PD-L1 positive non-small-cell lung cancer. *The New England Journal of Medicine, 375*, 1823–1833. https://doi.org/10.1056/NEJMoa1606774.

Rini, B. I., Plimack, E. R., Stus, V., Gafanov, R., Hawkins, R., Nosov, D., et al. (2019). Pembrolizumab plus axitinib versus sunitinib for advanced renal-cell carcinoma. *The New England Journal of Medicine, 380*(12), 1116–1127. https://doi.org/10.1056/NEJMoa1816714. Epub 2019 Feb 16. PMID: 30779529.

Robert, C., Long, G. V., Brady, B., Dutriaux, C., Maio, M., Mortier, L., et al. (2015). Nivolumab in previously untreated melanoma without BRAF mutation. *The New England Journal of Medicine, 372*, 320–330. https://doi.org/10.1056/NEJMoa1412082.

Robinson, J. P., & Roederer, M. (2015). History of science. Flow cytometry strikes gold. *Science, 350*, 739–740. https://doi.org/10.1126/science.aad6770.

Roussel, H., De Guillebon, E., Biard, L., Mandavit, M., Gibault, L., Fabre, E., et al. (2017). Composite biomarkers defined by multiparametric immunofluorescence analysis identify ALK-positive adenocarcinoma as a potential target for immunotherapy. *OncoImmunology, 6*, e1286437. https://doi.org/10.1080/2162402X.2017.1286437.

Schadendorf, D., et al. (2013). Pooled analysis of long-term survival data from phase II and phase III trials of ipilimumab in metastatic or locally advanced, unresectable melanoma [abstract]. *European Cancer Congress, 2013*, LBA24.

Shibru, B., Fey, K., Fricke, S., Blaudszun, A. R., Fürst, F., Weise, M., et al. (2021). Detection of immune checkpoint receptors—A current challenge in clinical flow cytometry. *Frontiers in Immunology, 1*(12), 694055. https://doi.org/10.3389/fimmu.2021.694055. PMID: 34276685; PMCID: PMC8281132.

Śledzińska, A., Menger, L., Bergerhoff, K., Peggs, K. S., & Quezada, S. A. (2015). Negative immune checkpoints on T lymphocytes and their relevance to cancer immunotherapy. *Molecular Oncology, 9*, 1936–1965. https://doi.org/10.1016/j.molonc.2015.10.008.

Spitzer, M. H., & Nolan, G. P. (2016). Mass cytometry: Single cells, many features. *Cell, 165*, 780–791. https://doi.org/10.1016/j.cell.2016.04.019.

Tanner, S. D., Baranov, V. I., Ornatsky, O. I., Bandura, D. R., & George, T. C. (2013). An introduction to mass cytometry: Fundamentals and applications. *Cancer Immunology, Immunotherapy, 62*, 955–965. https://doi.org/10.1007/s00262-013-1416-8.

Tognarelli, S., Jacobs, B., Staiger, N., & Ullrich, E. (2016). Flow cytometry-based assay for the monitoring of NK cell functions. *Journal of Visualized Experiments*. https://doi.org/10.3791/54615. doi: 10.3791/54615.

Wolchok, J. D., Chiarion-Sileni, V., Gonzalez, R., Rutkowski, P., Grob, J. J., Cowey, C. L., et al. (2017). Overall survival with combined nivolumab and ipilimumab in advanced melanoma. *New England Journal of Medicine, 377*(14), 1345–1356. https://doi.org/10.1056/NEJMoa1709684. Epub 2017 Sep 11. Erratum in: N Engl J Med. 2018 Nov 29;379(22):2185. PMID: 28889792; PMCID: PMC5706778.

Wu, C. P., Wu, P., Zhao, H. F., Liu, W. L., & Li, W. P. (2018). Clinical applications of and challenges in single-cell analysis of circulating tumor cells. *DNA and Cell Biology, 37*, 78–89. https://doi.org/10.1089/dna.2017.3981.

Yuan, J., Hegde, P. S., Clynes, R., Foukas, P. G., Harari, A., Kleen, T. O., et al. (2016). Novel technologies and emerging biomarkers for personalized cancer immunotherapy. *Journal for Immunotherapy of Cancer, 4*, 3. https://doi.org/10.1186/s40425-016-0107-3.

CHAPTER THREE

Multiparametric analysis of tumor infiltrating lymphocytes in solid tumors

Rebecca Borella[a,†], Annamaria Paolini[a,†], Beatrice Aramini[b], Lara Gibellini[a], Valentina Masciale[a], Domenico Lo Tartaro[a], Massimo Dominici[a], Sara De Biasi[a,*,‡], and Andrea Cossarizza[a,c,‡]

[a]Department of Medical and Surgical Sciences for Children & Adults, University of Modena and Reggio Emilia, Modena, Italy
[b]Division of Thoracic Surgery, Department of Medical and Surgical Sciences—DIMEC of the Alma Mater Studiorum—University of Bologna and G.B. Morgagni—L. Pierantoni Hospital, Forlì, Italy
[c]Istituto Nazionale per le Ricerche Cardiovascolari, Bologna, Italy
*Corresponding author: e-mail address: sara.debiasi@unimore.it

Contents

Abstract

The use of single-cell technologies in characterizing the interactions between immune and cancer cells is in continuous expansion. Indeed, the combination of different single-cell approaches enables the definition of novel phenotypic and functional aspects of the immune cells infiltrating the tumor microenvironment (TME). This

† These authors have contributed equally to this work.
‡ These authors share senior authorship.

Methods in Cell Biology, Volume 195
ISSN 0091-679X
https://doi.org/10.1016/bs.mcb.2023.03.006

approach is promoting the discovery of relevant and reliable predictive biomarkers, along with the development of new promising treatments. In this chapter, we describe the main subsets of tumor-infiltrating lymphocytes from a phenotypical and functional point of view and discuss the use of single-cell technologies used to characterize these cell populations within TME.

1. Introduction

A successful antitumor response by the immune system requires the presence and activation of a huge number of actors, including CD8$^+$ cytotoxic T cells (CTLs), CD4$^+$ T cells, B cells, and innate lymphoid cells (Paijens, Vledder, de Bruyn, & Nijman, 2021). The function of the immune system is thus to contain tumor development and progression by exerting at least three different roles: (i) the suppression of viral infection that can eventually lead to develop tumor of viral etiology; (ii) the prevention or resolution of inflammatory processes that promote tumorigenesis; (iii) the clearance of tumor cells that in some tissues can be identified by the expression of unique tumor antigens (Schreiber, Old, & Smyth, 2011). This immunosurveillance is accomplished by the infiltration of adaptive and innate immune cells into the TME (Schreiber et al., 2011; Seager, Hajal, Spill, Kamm, & Zaman, 2017). Innate immune cells are composed by natural killer (NK) cells, eosinophils, basophils, and elements with phagocytic activity that include mast cells, neutrophils, monocytes, macrophages, and dendritic cells (DCs), whose metabolism plays a key role in modulating their activities in health or disease (Borella et al., 2022; Gibellini et al., 2020; Zanini et al., 2021).

The adaptive immune system is formed by B lymphocytes that typically not only produce antibodies and play a major role in humoral immune responses, but can also regulate the activity of other cells, and by T lymphocytes that are responsible for cell-mediated immune responses, also including the production and release of a variety of soluble molecules. Innate immune cells participate in tumor suppression either by directly killing tumor cells or triggering adaptive immune response (Corrales, Matson, Flood, Spranger, & Gajewski, 2017; Demaria et al., 2019; Woo, Corrales, & Gajewski, 2015).

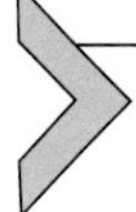

2. Tumor-infiltrating lymphocytes

2.1 Heterogeneous composition of the tumor microenvironment

Tumor microenvironments (TMEs) are heterogeneous and dynamic contexts which differ for composition, functional status, and spatial distribution of immune cells. TMEs are characterized by intricate crosstalk among tumor cells, tumor stromal cells (including stromal fibroblasts and endothelial cells) and immune elements, among which tumor infiltrating lymphocytes (TILs) can be found. TILs are lymphocytes that migrate from the blood to the tumor across the endothelial barrier (Bouzin, Brouet, De Vriese, Dewever, & Feron, 2007). Their presence has been associated with an improved clinical outcome, that is influenced by the phenotype and function of T cells involved [CTLs or regulatory T cells (Tregs)] as well as by the intratumor localization of TILs (Fridman, Pages, Sautes-Fridman, & Galon, 2012; Tsakiroglou et al., 2020).

TME is enriched in non-cellular components of extracellular matrix, and is also rich of collagen, fibronectin, hyaluronan and laminin (Baghban et al., 2020). In this context, tumor cells control the function of cellular and non-cellular components through a complex intercellular signaling composed of cytokines, chemokines, growth factors, inflammatory mediators, enzymes, and other newly discovered mechanisms of interaction such as exosomes, cell-free DNA, and apoptotic bodies (Baghban et al., 2020). Two types of TMEs have been identified: (i) infiltrated-excluded TMEs that are populated by immune cells mainly along the tumor margins and are relatively poor of CTLs in the tumor core. Here, CTLs display a low expression of activation and cytotoxic molecules such as granzyme B (GZMB) and interferon-γ (IFN-γ); (ii) infiltrated-inflamed TMEs that are defined by an important immune infiltration among neoplastic cells and a high frequency of CTLs expressing GZMB, IFN-γ and programmed cell death protein 1 (PD-1) (Binnewies et al., 2018). In tumors characterized by this TME structure, compartments exist that resemble tertiary lymphoid structures (TLSs). These sites are usually located both at the margin of the tumor and in the stroma, and are enriched in naïve and activated T cells, Tregs, B cells and DCs (Binnewies et al., 2018).

2.2 TILs isolation from solid tumors

In the field of immune oncology, the characterization of TME provides important insight into the assessment of therapeutics against tumor progression. Since a critical role in anti-tumor immune response is played by TILs, their isolation from tumors represents a desirable first approach before downstream analysis such as flow cytometry, functional characterization, and analysis of gene expression. Different protocols exist for TILs isolation from excised tumors. TILs are commonly purified using Ficoll-Paque density gradient centrifugation, which is broadly adaptable for human samples, although other silica-based density gradient media can be used for centrifugation, such as Percoll.

For tumor dissociation there are two main approaches, mechanical or enzymatic, and a combination of both is also feasible. Mechanical dissociation of tumor to obtain single-cell suspensions consists in excising tumor using scissors and forceps, and mincing tumor into small pieces (approximately 2 mm). Enzymatic digestion is performed by treating tumor fragments with a digest solution containing a cocktail of enzymes such as collagenase and deoxyribonuclease. During this procedure it is crucial to preserve the highest viability of the TILs in order to investigate adequately their function and phenotype. In particular, the type of enzymatic treatment, in terms of enzyme choice and duration of dissociation must be carefully selected and optimized based on the type of tumor, in order to obtain the optimal viable cell yield and to preserve TILs surface markers (Leelatian et al., 2017; Tan & Lei, 2019).

2.3 The use of single-cell technologies to characterize TILs in TME

Single-cell analytical approaches represent a revolutionary resource for investigating complex biological processes in health and diseases. During the last decades, the wide use of high parametric cytometry has provided the classification of several cell populations and the discovery of new cellular subsets that form the immunological landscape. The phenotypic and functional profile of a cell is highly heterogeneous, so a large number of markers must be used to fully characterize it. Moreover, multidimensionality as well as high resolution are required to measure poorly expressed molecules. Ongoing improvements extend the applicability of single-cell technologies to several areas of study. In cancer research, flow cytometry is critical for monitoring immune changes that can occur both in peripheral blood and tumor microenvironment. The aforementioned power of this technique enables a reliable identification of rare cell populations that are present in a population of millions of elements, including those that are specific for antigens expressed by tumor cells.

As previously mentioned, TME composition is very heterogeneous as well as its contribution to the control of immune cells function inside the tumor (Aramini et al., 2022). Therefore, to identify and precisely characterize the phenotype of TILs in such environment, the use and eventual combination of single-cell technologies is highly recommended (Gibellini et al., 2020). Currently, flow cytometry allows for the simultaneous identification of up to 40 markers at the single cell level, enabling to measure a number of physical and chemical properties of thousands of cells or particles in a very short time, and generating massive datasets (Brummelman et al., 2019; Chattopadhyay, Winters, Lomas 3rd, Laino, & Woods, 2019; Liechti et al., 2021). Imaging techniques such as immunofluorescence microscopy, electron microscopy and imaging mass cytometry (IMC) enable the characterization of single cells and of the heterogeneous environment in which they exert their function and allow to consider also the spatial distribution and complexity, together with the study of cell-cell interactions and of intercellular networks within the tissue (Chang et al., 2017; Giesen et al., 2014; Schapiro et al., 2017; Wagner et al., 2019).

Multiplexed ion beam imaging (MIBI) is an alternative approach to IMC (Chang et al., 2017), while histo-cytometry allows for the visualization and quantification of phenotypically complex cell populations (characterized by the expression of multiple markers) adding a positional analysis directly in tissue sections obtaining results similar to those of flow cytometry, with additional gathering of cellular positional information (Gerner, Kastenmuller, Ifrim, Kabat, & Germain, 2012). Co-detection by indexing (CODEX) is a technology that employs oligonucleotide-conjugated antibodies (Goltsev et al., 2018) enabling a deep view into the single-cell spatial relationships in tissues and is intended to spur discovery in developmental biology, disease, and therapeutic design (Black et al., 2021; Kennedy-Darling et al., 2021).

Fluorescence-based technologies follow different approaches for the detection of fluorescence, i.e., the use of individual filters and photomultipliers, or detectors for spectral analysis. Well-designed experiments require adequate panel design, selection of markers and fluorochromes, titration of antibodies, optimized use of reagents such as fixation/permeabilization buffers, depending on whether intracellular antigens localize to cytosol or nucleus, and, finally, quality checks and validation concepts. In principle, flow cytometry analysis is versatile, high-throughput and affordable, and several protocols providing instructions to perform high-quality flow cytometry analysis, including panel design, have been published in main journals (Brummelman et al., 2019). Moreover, constantly updated guidelines for a correct use of cytometry exist and are freely available (Cossarizza et al., 2017, 2019, 2021).

The data resulting from every single sample acquisition are saved as a Flow Cytometry Standard (FCS) file, where all the information concerning physical parameters and fluorescence intensity obtained for each probe on each individual cell are collected. FCS files can be analyzed using several commercial software packages, mostly based on sequential "gating" or selection of populations of interest. The quantification of the amount of molecules present in a single cell, as well as their absolute count also possible, for example, by referring to standardized labeled beads. Automated gating algorithms can also be used, some of which allow an easy representation of the data by unsupervised clustering and dimension reduction approaches (Maecker & Harari, 2015).

2.4 Phenotypic and functional characterization of TILs

The infiltration of T cells into the TME plays a critical role for immune-mediated control of cancer; for this reason, a variety of approaches aim to restore and/or boost T-cell function in cancer patients (Hinrichs & Rosenberg, 2014; Sadelain, Riviere, & Riddell, 2017). $CD8^+$ T cells are key players in this context, thanks to their cytotoxicity and capacity of releasing cytotoxic molecules and different types of cytokines (Klebanoff, Gattinoni, & Restifo, 2006). Tumor specific $CD8^+$ T cells are primed in tumor-draining lymph nodes (LN) and then migrate to the tumor. Here they can recognize tumor-associated antigens, that can include either overexpressed antigen or neoantigens derived from tumor mutations (Finn, 2017).

Among the main actors of T cell immunity, central memory T cells (TCMs), effector memory T cells (TEMs) and resident memory T cells (TRMs) play a main role (Mahnke, Brodie, Sallusto, Roederer, & Lugli, 2013; Mueller & Mackay, 2016) (Fig. 1). These three lymphocyte subsets can respond immediately to a given stimulus. Each subset is identified by the expression of specific surface molecules and specific localization within tissues, lymph nodes or peripheral blood (Marceaux, Weeden, Gordon, & Asselin-Labat, 2021). TCMs express CD27, CD28 and CC-chemokine receptor (CCR) 7, but not CD45RA, whereas TEMs are negative for all these markers (Martin & Badovinac, 2018). An intermediate differentiation phenotype characterizes an additional memory T-cell subset named "transitional memory T cells" (TTMs) that have already lost CCR7 and CD62L expression but are still positive for CD28. This profile makes them more differentiated than TCMs, but not as fully differentiated as TEMs (Fritsch et al., 2005; Okada, Kondo, Matsuki, Takata, & Takiguchi, 2008). In addition, TCM populations express CD62L, and both TCM and TEM subsets are positive for CD45R0.

CD27 and CD28 are costimulatory molecules expressed on most naïve $CD4^+$ and $CD8^+$ peripheral T cells (Martin & Badovinac, 2018; Powell Jr., Dudley, Robbins, & Rosenberg, 2005). The interaction of these receptor with their ligands (CD70 for CD27, CD80 and CD86 for CD28) amplifies T cell receptor (TCR)-mediated T-cell activation and

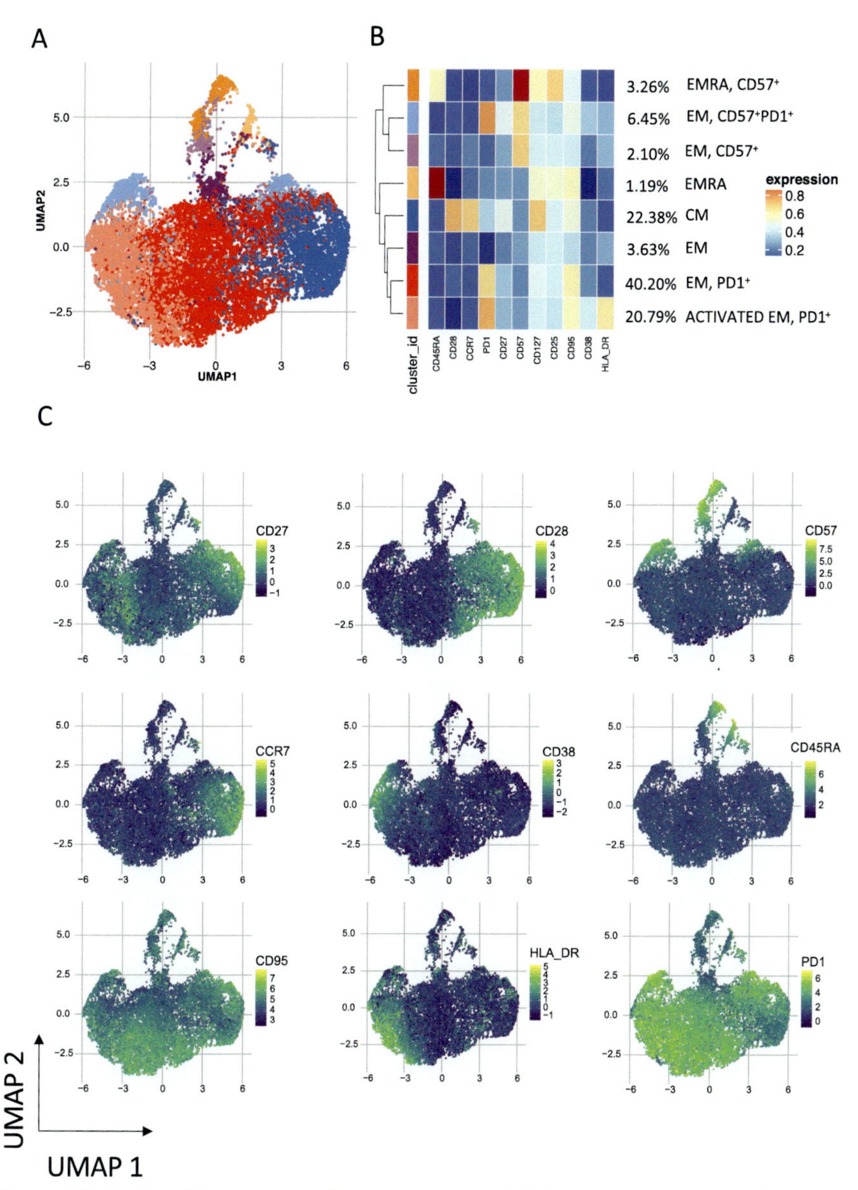

Fig. 1 TILs isolated from non-small cell lung cancer (NSCLC) patients are mostly TEMs. Unsupervised analysis and characterization of TILs isolated from biopsies obtained by 10 patients undergoing surgery. TILs were thawed and stained with Duraclone IMT besides viability marker and directly conjugated mAbs: CD38, HLA-DR, CD95, CD25, CD127. FCS files were analyzed, compensated, and transformed in FlowJo and $CD8^+$ TILs were exported per each patient and analyzed by using Cytometry dATa anALYSis Tools (CATALYST). (A) Uniform Manifold Approximation and Projection

(Continued)

proliferation (Hintzen, de Jong, Lens, & van Lier, 1994). A prolonged TCR stimulation results in the downregulation of CD27 and CD28. As a consequence, the loss of this expression defines a population of late-stage effector cells that have lost the ability to proliferate and have an increased cytotoxic potential (Powell Jr. et al., 2005; Romero et al., 2007). The expression of CCR7 and CD62L on TCMs facilitates homing to secondary lymphoid organs, while TEMs express integrins and chemokine receptors that drive them to the sites of inflammation (Forster, Davalos-Misslitz, & Rot, 2008; Martin & Badovinac, 2018). CD45RA and CD45R0 are two isoforms of CD45 that regulate T cell activation. The selective expression of one isoform depends on the stage of T-cell maturation, activation, differentiation and age of the donor. CD45RA is characteristic of naïve T cells, whereas CD45R0 is expressed on activated and memory T cells (Carrasco, Godelaine, Van Pel, Boon, & van der Bruggen, 2006; Cossarizza et al., 1996; Mahnke et al., 2013; Marceaux et al., 2021).

Localization is also an important feature for the classification of T memory subsets. TCMs can be found mostly in secondary lymphoid organs and in circulation, while TEMs circulate in the blood and transiently traffic to tissues (Kumar, Connors, & Farber, 2018; Marceaux et al., 2021). Tissue resident memory T cells (TRMs) are effector memory cells that do not circulate, but reside permanently within the tissue. They are marked by the expression of specific proteins that can differ depending on tissue of residence and include integrins CD103, CD49a which allow cells to enter into the tissue by binding respectively e-cadherin expressed on normal host epithelial cells, and collagenase type IV required for the normal turnover of basement membranes (Cheuk et al., 2017; Molodtsov & Turk, 2018; Richter & Topham, 2007; Stetler-Stevenson, 1990). Another molecule

Fig. 1—Cont'd (UMAP) representation of the landscape of TILs. (B) Heatmap representing different clusters of $CD8^+$ TILs identified by FlowSOM, with relative identity and percentages in NSCLC patients. The colors in the heatmap represent the median of the arcsine, 0–1 transformed marker expression calculated over cells from all the samples, varying from blue for lower expression to red for higher expression. The dendrogram on the left represents the hierarchical similarity between the metaclusters (metric: Euclidean distance; linkage: average). Each cluster has a unique color assigned (bar on the left). TILs isolated from cancer patients are mostly TEMs expressing high level of PD-1 and CD95. A small population represents terminally differentiated effector memory T cells (EMRA) expressing CD57 and PD-1. A small fraction of effector memory (EM) is activated ($CD38^+$, $HLA\text{-}DR^+$) and expresses CD27, indicating the presence of functional effector T cell pool. (C) UMAP graphs colored according to the expression each single marker out of 9 that were used for clustering of TILs.

expressed on TRMs is CD69, a marker of T cell activation which promotes TRM retention and residency within tissue by binding sphingosine-1-phosphate (S1P)-receptor 1 (S1PR1) on T cells. This interferes with the ability to sense the S1P gradient that is essential to leave the tissues (Baeyens, Fang, Chen, & Schwab, 2015; Mackay et al., 2013; Molodtsov & Turk, 2018). TRMs also express chemokine receptors such as CXC-chemokine receptor 3 (CXCR) 3 which is required for appropriate localization in the skin or to the lung epithelium, but negative for CCR7 and CD62L (Martin & Badovinac, 2018). An example of the identification of TRMs by flow cytometry is shown in Fig. 2.

TILs are effector T cells exhibiting different phenotypes and functions. In particular, $CD8^{+}$ T cells with a cytotoxic profile and $CD4^{+}$ Tregs have a central role in cancer immunity (Maibach, Sadozai, Seyed Jafari, Hunger, & Schenk, 2020). Tumor specific $CD8^{+}$ T cells that are critically involved in immune-mediated control of cancer exhibit a TRM phenotype when they migrate within the tumor, after their previous priming in lymph nodes (Dumauthioz, Labiano, & Romero, 2018; Molodtsov & Turk, 2018). $CD8^{+}$ TRMs are defined based on their long-term persistence in tissues without re-entering the bloodstream (Molodtsov & Turk, 2018). $CD8^{+}$ TRMs express not only typical markers of residency, such as CD103 and CD49a, but also chemokine receptors such as CXCR6 which plays a critical role in resident memory T cell-mediated immunosurveillance. CXCR6 is highly expressed in tumor-specific T cells that are resident in tissues but not by those in circulation (Muthuswamy et al., 2021). $CD8^{+}$ TILs also display an activated phenotype expressing molecules such as CD69, HLA-DR and CD38. HLA-DR (i.e., the human leukocyte antigen) is a class II MHC molecule, normally expressed on professional antigen presenting cells (APCs), and is also associated with increased IFN-γ production. CTLs upregulate this molecule 24–48 hours after the first activation, when CD69 has already been upregulated. The increase of HLA-DR at CTL surface is required to boost an effective immune response (Saraiva et al., 2021). CD38 is a surface receptor involved in different functions such as lymphocyte activation, proliferation and survival, often depending on the cell type (Malavasi et al., 2008; March et al., 2007; Ortolani, Forti, Radin, Cibin, & Cossarizza, 1993; Wo et al., 2019). The co-expression of CD38 and HLA-DR is considered a feature to identify recently activated tumor-antigen specific T cells (Kovacsovics-Bankowski et al., 2014).

TILs do not express the selectin CD62L as they belong to the tissue resident subpopulations. CCR5 and CCR6 are chemokine receptors that play

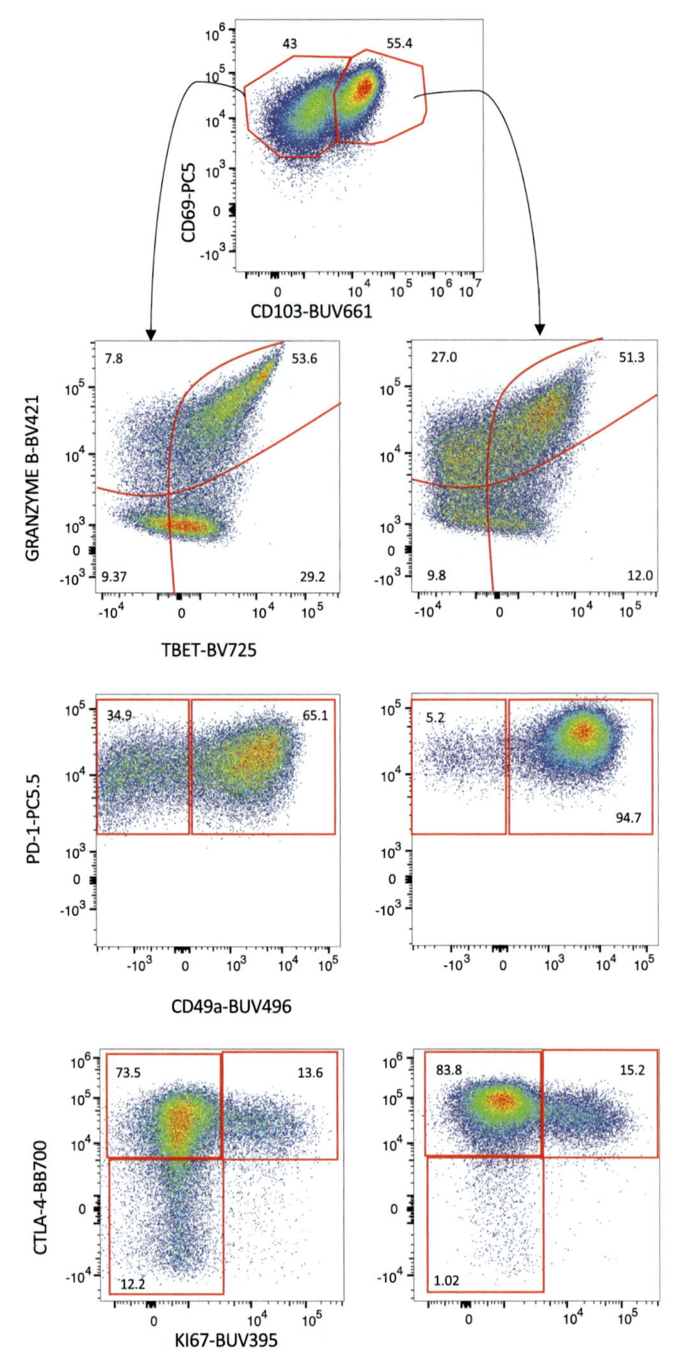

Fig. 2 Identification of TRMs by manual gating, using a 21-color panel. Example of phenotypic characterization of TILs isolated from NSCLC. First, $CD45^{+}CD3^{+}CD8^{+}$ viable cells have been electronically gated. In this gate, two populations can be identified as $CD103^{+}CD69^{+}$ or $CD103^{-}CD69^{+}$ T cells. The dot plots show that different expression of GZMB, T-box transcription factor TBX21 (TBET), PD-1, CD49a, CTLA-4 and Ki-67 exist between the two subsets.

a crucial role in the selective T cell recruitment within the TME (Liu et al., 2015). CCR5 is mostly expressed on $CD8^+$ T cells and can bind various ligands such as CC-chemokine ligand (CCL) 3, CCL4, CCL5, CCL7, CCL11, CCL14, CCL16. Conversely, CCR6 is mostly expressed on Tregs and regulates their recruitment by binding CCL20 (Liu et al., 2015). Ki-67 is a nuclear protein involved in the regulation of cell proliferation and it is expressed during all active phases of the cell cycle (G1, S, G2 and M), but not by resting cells (Li, Jiang, Chen, & Zheng, 2015; Shirendeb et al., 2009). Since it reflects the ability of cells to proliferate, it has been extensively used as a prognostic indicator in cancer (Wang et al., 2021). Interestingly, the presence of Ki-67 in a cell can define two different scenarios: if expressed by tumor cells, it has been associated to poor prognosis (Brown & Gatter, 2002; Kloppel, Perren, & Heitz, 2004), whereas, if expressed by TILs, it indicates a favorable clinical outcome (Wang et al., 2021). Therefore, in order to understand its meaning, its identification has to be accompanied by the study of several other molecules.

2.5 The role of regulatory T cells

Tregs are highly abundant within the TME (Roychoudhuri, Eil, & Restifo, 2015). Tregs are $CD4^+$ T cells expressing high levels of the interleukin (IL)-2 receptor α-chain (IL-2Rα), also known as CD25, and lacking the CD127, the α-chain of the IL-7 receptor (Liu et al., 2006; Ohue & Nishikawa, 2019). Forkhead box P3 (Foxp3), a member of the forkhead/winged-helix family of transcriptional regulators, is a master regulatory gene involved in the generation, maintenance, and functions of Tregs (Ohue & Nishikawa, 2019). This nuclear factor controls the immunosuppressive function of Tregs which is maintained through its continuous expression (Li et al., 2020). The main role of Tregs is the maintenance of the homeostasis that leads to the inhibition of effector T cell functions in the tumor. Tregs can exert this role in different ways: (i) expressing cell-surface inhibitors such as cytotoxic T lymphocyte–associated protein 4 (CTLA-4, or CD152) (Wing et al., 2008), (ii) producing inhibitory cytokines such as IL-10 (Larmonier et al., 2007), transforming growth factor (TGF)-β (Chen et al., 2005), or IL-35 (Collison et al., 2007), (iii) causing the depletion of IL-2 by overexpressing IL-2Rα (Alvisi et al., 2020), or (iv) even with direct cytotoxicity (Grossman et al., 2004). Within the tumor, Tregs display a T cell effector phenotype ($CD45RA^-$, $CD45R0^+$, $CCR7^-$,$CD27^-$, $CD28^-$) (Chen & Oppenheim, 2011; Plitas et al., 2016) and acquire an enhanced suppressive capacity. They express higher amounts of suppressive molecules such as CTLA-4, inducible T-cell costimulator (ICOS) and IL-10 (Alvisi et al., 2020) compared to the counterparts present in the blood and in adjacent tumor-free tissues (De Simone et al., 2016; Plitas et al., 2016).

Treg infiltration and abundance within the tumor microenvironment do not correlate with better prognosis (Tanaka & Sakaguchi, 2017). In cancer treatment, systemic depletion of Tregs has been taken into consideration as a therapeutic approach. However, depletion or inhibition of Tregs often results in a detrimental loss of tissue homeostasis that leads to severe autoimmunity or allergy. For these reasons, the identification and targeting of specific molecules that are exclusively present in tumor-associated Tregs would be a more beneficial strategy that could avoid side effects (Alvisi et al., 2020; Colombo & Piconese, 2007; Dannull et al., 2005; Luo, Liao, Dadi, Toure, & Li, 2016; Wei et al., 2005). Two examples can be interferon regulatory factor 4 (IRF4) and CCR8, whose high expression on Tregs in TME strongly correlates with poor prognosis and decreased survival in multiple human cancers (Alvisi et al., 2020; Kidani et al., 2022; Li, Wang, et al., 2020; Plitas et al., 2016; Swatler et al., 2022).

2.6 Cytokine and chemokine signals regulate TILs recruitment and function within TME

Cytokines and chemokines modulate cellular compositions of the TME. They are released by cancer, stromal, and immune cells and, upon binding to their cognate receptors, regulate recruitment, activation, apoptosis of target cells, among others. Chemokines mainly regulate immune cell trafficking by providing the chemotactic signals. Cytokines such as IL-6 and tumor necrosis factor (TNF) enhance the endothelium adhesion activity by promoting adhesion molecules' expression on tumor vessels. CXC-chemokine ligand (CXCL) 9, CXCL10, and CXCL11, secreted by tumor and stromal cells, regulate chemotaxis of T cells after extravasation. These ligands and other chemokines such as CCL5, CXCL16, CCL21, and CCL27 positively correlate with T cell recruitment and abundance within solid tumors. Within TME, there are also negative signals that hamper T cell infiltration. These include IL-35 released by Tregs, macrophages, or B cells which promote T cell exhaustion by regulating the expression of several inhibitory receptors. TGF-β is secreted by stromal cells, can induce angiogenic factors like vascular endothelial growth factor (VEGF) and recruits immune suppressive cells such as Tregs and myeloid-derived suppressor cells (MDSCs) present within the tumor. After binding specific membrane receptors, TGF-β inhibits $CD8^+$ T cells and DCs, activates Tregs and stimulates chemotaxis of polymorphonuclear cells and monocytes (Li, Wang, et al., 2020; Zhang, Guan, & Jiang, 2020). IL-2 family members (i.e., IL-7, IL-15, IL-21) play a crucial role for $CD8^+$ T-cell expansion in the hostile TME.

Other chemokines trigger apoptosis of T cells infiltrating the tumor. For example, CCL5 secreted by TILs induces non-classic apoptosis of T cells through the binding of CCR5. When secreted by tumor-infiltrating $CD4^+$ T cells, as reported in case of gastric cancer, CCL5 also facilitates $CD8^+$ T cell apoptosis mediated by Fas-Fas ligand signaling (Sugasawa et al., 2008; Zhang et al., 2020).

The ability of cells to produce specific cytokines is a relevant feature that defines their function. TILs can secrete IFN-γ that improves antigen presentation by DCs and macrophages and enhances the production of costimulatory molecules and cytokines that are necessary for T cell activation (Gao et al., 2016; Park et al., 2017). Additionally, IFN-γ promotes apoptosis of tumor cells and immune recognition by the induction of both MHC class I molecules and signal transducer and activator of transcription 1 (STAT1)-associated cyclin-dependent kinase (Burke & Young, 2019). TNF is a pleiotropic cytokine involved in a range of physiological processes controlling inflammation, anti-tumor responses, and immune system homeostasis (Mehta, Gracias, & Croft, 2018; Montfort et al., 2019). TNF can exhibit both anti- and pro-tumor characteristics being able to modulate the activation, function, and survival of leukocytes during cancer progression, but also to alter the phenotype of cancer cells making them less visible to T cells, maintain inflammation and to favor the expression of immune inhibitory molecules (Montfort et al., 2019).

Finally, $CD8^+$ CTLs can also exert their cytotoxic activity by producing proteases such as GZMB (Cullen, Brunet, & Martin, 2010). Cells expressing GZMB exert a more effective and rapid killing due to their ability to target cell apoptosis of the target cell by caspases (Cullen et al., 2010). Tregs can also produce GZMB to kill effector cells, as well as inhibitory cytokines including IL-10 and TGF-β (Li, Wang, et al., 2020).

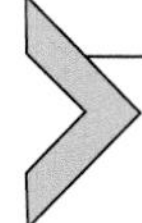

3. Immune checkpoint proteins (ICPs) expression and tumor immune surveillance

Tumor cells suppress immune surveillance functions in different ways to avoid being attacked by the immune system. They downregulate the activity of stimulatory immunoreceptors and upregulate the activity of inhibitory immunoreceptors. In the last decades, different inhibitory immunoreceptors have been identified and studied and have been defined immune checkpoint proteins (ICPs) including programmed death protein 1 (PD-1, or CD279), CD39, CTLA-4, T cell immunoreceptor with immunoglobulin and ITIM domain (TIGIT), T cell immunoglobulin and mucin domain containing molecule-3 (TIM-3), lymphocyte activation gene 3 (LAG-3), and V-domain Ig suppressor of T cell activation (VISTA). These proteins, when expressed on the cell surface, are crucial for self-tolerance as they have to inhibit indiscriminate attacks of immune cells against self-components (He & Xu, 2020; Pardoll, 2012) (Fig. 3). However, their dysregulated expression impairs T cell activation and promotes mechanisms of tumor immune resistance, including an uncontrolled cancer cell proliferation (Pardoll, 2012; Stirling et al., 2022).

PD-1 is commonly expressed on the surface of activated T cells, B cells, monocytes, DCs, Tregs and natural killer T cells (NKT) (De Biasi et al., 2020; Gibellini, De Biasi, Paolini, et al., 2020; Keir, Butte, Freeman, & Sharpe, 2008). In the TME, PD-1 is expressed on a large proportion of TILs including Tregs and its ligand, named PD-L1, normally expressed on APCs (Jiang, Chen, Nie, & Yuan, 2019), is upregulated by tumor cells. The binding of PD-1 to PD-L1 downregulates several T cell functions, such as proliferation, cytokine secretion, cytotoxicity and also tumor infiltration by TILs (Jiang et al., 2019; Liu et al., 2021). This immunosuppression defines a profile of exhaustion: an increased PD-1 expression on $CD8^+$ TILs reflect an anergic state, indicating a loss of CTL function (Jiang et al., 2019) that enables tumor cells to evade immune surveillance (Jiang et al., 2019; Liu et al., 2021). On the other hand, PD-1 inhibition of Tregs potentiate the activation and immunosuppression of this subset (Ohue & Nishikawa, 2019).

CD39 is an ectonucleotidase present on the surface of activated $CD4^+$ and $CD8^+$ T cells (Kortekaas et al., 2020). Within TME, the chronic TCR stimulation and the presence of cytokines such as IL-6, IL-27 and TGF-β upregulates CD39 on T cells (Canale et al., 2018; Duhen et al., 2018). CD39 hydrolyses extracellular adenosine triphosphate (eATP) to adenosine. eATP is one of major biochemical constituents of the TME and can promote cancer cell growth or trigger cancer cell death. Moreover, it can be involved in immune cell recruitment and activation but can also promote immunosuppression when it is degraded to adenosine. Actually, its degradation by CD39 reduces effector $CD8^+$ T cell responses and favours Tregs expansion, thus causing immunosuppression (Antonioli, Pacher, Vizi, & Hasko, 2013; Junger, 2011; Timperi & Barnaba, 2021;

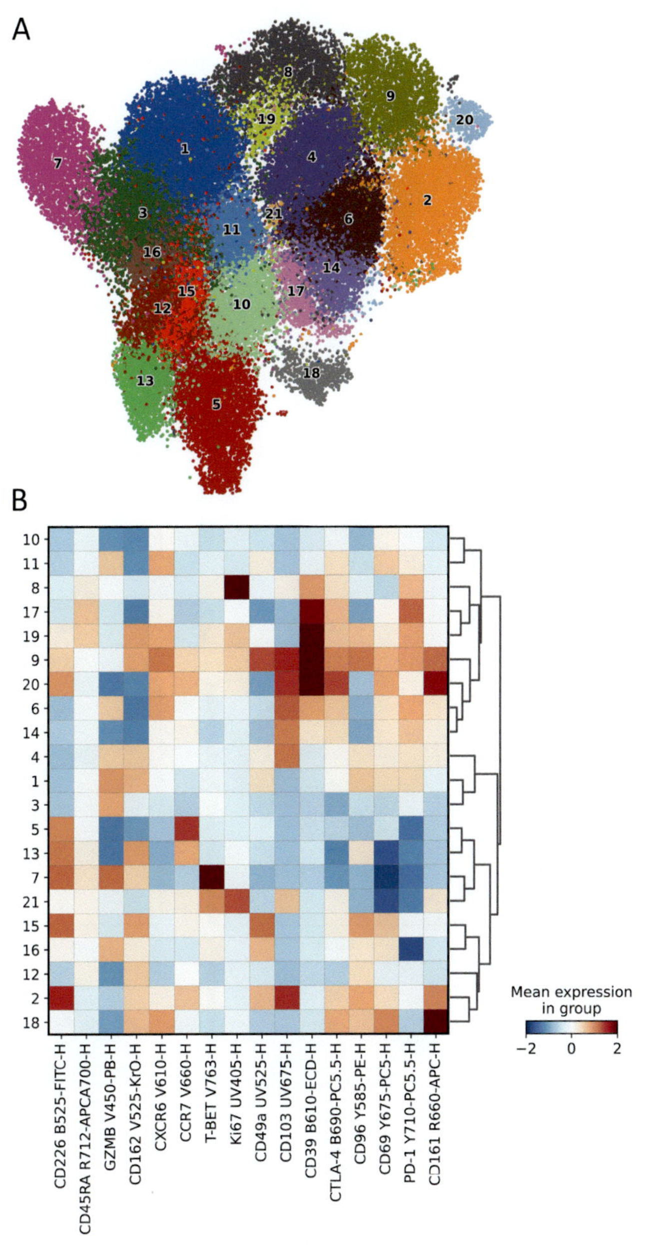

Fig. 3 Unsupervised cluster analysis of CD8$^+$ TILs isolated from a NSCLC patient undergoing surgical treatment. Example of 21-color panel used to identify different subpopulations of TILs, including TRMs, along with different immune checkpoint receptors, activation and homing molecules. The FCS file was analyzed, compensated, and transformed in FlowJo and CD8$^+$ TILs were exported and analyzed by a custom-made pipeline of PhenoGraph (available at: http://github.com/luglilab/Cytophenograph). (A) UMAP representation of the landscape of TILs. (B) Heatmap representing different clusters of CD8$^+$ TILs identified by PhenoGraph. The colors in the heatmap represent the median of the arcsine, −2+2 transformed marker expression calculated over cells from all the samples, varying from blue for lower expression to red for higher expression.

(Continued)

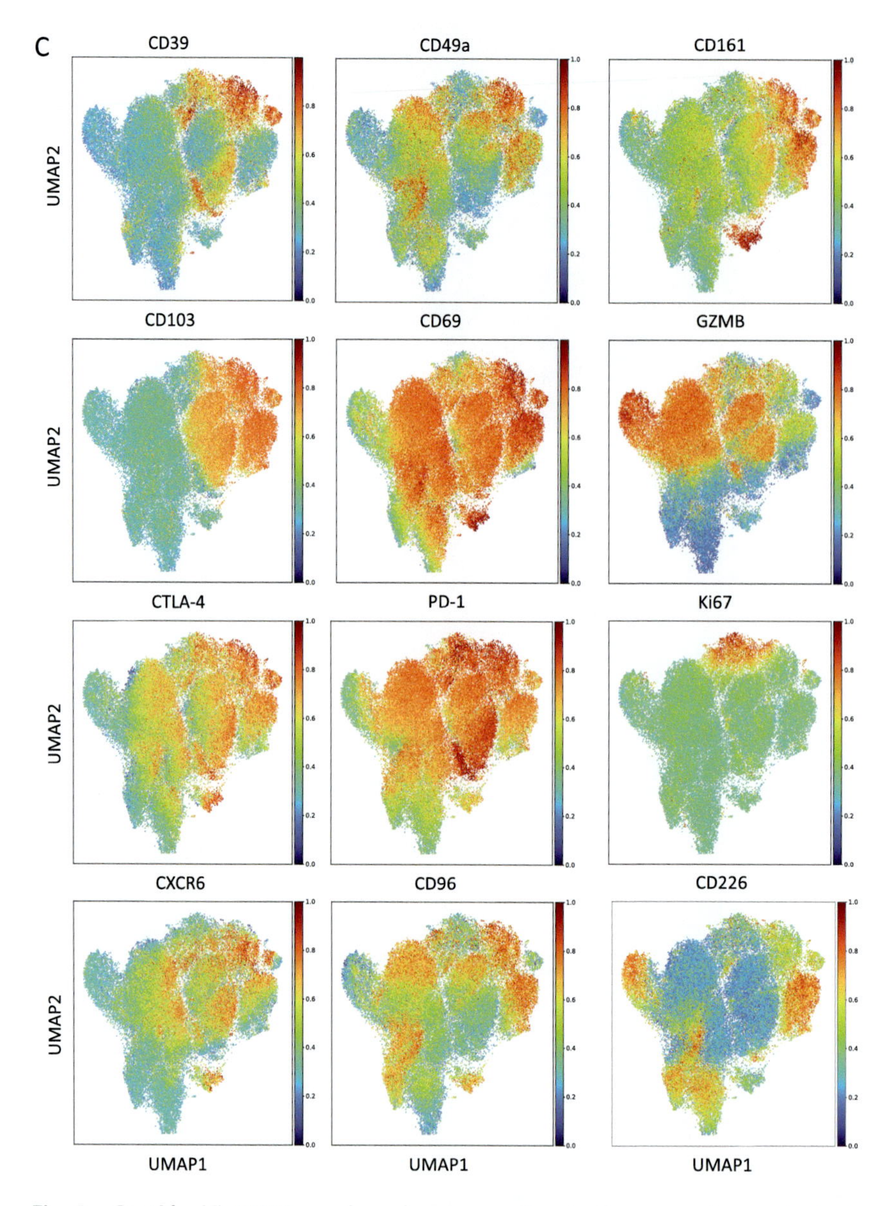

Fig. 3—Cont'd (C) UMAP graphs colored according to the expression each single marker out of 12 that were used for clustering of TILs.

Vultaggio-Poma, Sarti, & Di Virgilio, 2020). Therefore, the expression of CD39 defines both a population of exhausted $CD8^+$ T cells and of immunosuppressive Tregs (Cai et al., 2016; Canale et al., 2018) and in cancer biology is considered to be a marker of poor outcome and disease progression (Cai, Wang, et al., 2016; Canale et al., 2018; Simoni et al., 2018).

CTLA-4 is a receptor expressed by both $CD4^+$ and $CD8^+$ T cells. It has a homologous structure as CD28; however, after triggering, the two receptors exert opposite actions on T cell activation (Gibellini, De Biasi, Porta, et al., 2020; Melichar, Nash, Lenzi, Platsoucas, & Freedman, 2000; Rowshanravan, Halliday, & Sansom, 2018). They both interact with CD80 and CD86 expressed on APCs, but CTLA-4 shows a higher binding affinity and avidity (Melichar et al., 2000; Rowshanravan et al., 2018). Interaction of CD80 and CD86 with CD28 serves as co-stimulation in conjunction with TCR signals, while their binding with CTLA-4 inhibits T cell responses. Within the TME, CTLA-4 mediates the suppressive function of Tregs and blocks $CD8^+$ effector T cell signaling causing exhaustion and promoting tumor immune escape (Rowshanravan et al., 2018).

TIGIT is an inhibitory receptor that participates in a complex regulatory network involving multiple receptors and ligands (Chauvin & Zarour, 2020). It is expressed by activated human $CD8^+$ and $CD4^+$ T cells, NK cells, Tregs and T helper (Th) cells (Joller et al., 2011; Stanietsky et al., 2009). In cancer, TIGIT is co-expressed with other immune checkpoints such as PD-1, TIM-3 and LAG-3, on exhausted $CD8^+$ T cell (Chauvin & Zarour, 2020). In addition, it is highly expressed on Tregs both in peripheral blood and tumor tissue (Anderson, Joller, & Kuchroo, 2016). TIGIT, as the other ICPs, negatively regulates anti-tumor responses inhibiting T cell effector function and enhancing immunosuppressive action of Tregs (Anderson et al., 2016). To confirm that, it was reported that TIGIT deficiency significantly delayed tumor growth in two different tumor models (Kurtulus et al., 2015).

TIM-3 is expressed on a variety of cell subsets, such as $CD4^+$ Th1 and $CD8^+$ CTLs, Tregs and on innate immune cells (Monney et al., 2002). Regarding $CD8^+$ TILs, the expression of TIM-3 and PD-1 can be used to characterize different populations that exhibit different functional phenotypes (Sakuishi et al., 2010). $TIM\text{-}3^+PD\text{-}1^+CD8^+$ TILs exhibit a severe dysfunctional and exhausted phenotype, $TIM\text{-}3^-PD\text{-}1^+CD8^+$ TILs exhibit a weak dysfunctional exhausted phenotype, whilst $TIM\text{-}3^-PD\text{-}1^-CD8^+$ TILs exhibit good effector functions (Kurtulus et al., 2015). A small subset of circulating Tregs expresses TIM-3 in the secondary lymphoid organs and peripheral blood while a significantly higher amount of $TIM\text{-}3^+$ Tregs is found among tumor-infiltrating Tregs (Banerjee et al., 2021; Gao et al., 2012; Sakuishi et al., 2013; Yan et al., 2013). TIM-3 expression on Tregs enhances cell activation and drives this subset to a more effector-like phenotype. This results in an increased suppressive activity which impairs T cell responses to tumors promoting tumor growth (Banerjee et al., 2021).

LAG-3 is upregulated on activated $CD4^+$ and $CD8^+$ T cells and on a subset of NK cells (Triebel et al., 1990). It is a negative regulator of T cell expansion (Workman et al., 2004; Workman & Vignali, 2003) as reported by experiments on LAG-3 deficient mouse where an uncontrolled T cell proliferation was observed (Anderson et al., 2016). LAG-3 is expressed on exhausted $CD8^+$ T cells (Maruhashi, Sugiura, Okazaki, & Okazaki, 2020) and even more expressed on regulatory T cells, (Huang et al., 2004) where it controls T cell homeostasis (Workman & Vignali, 2005). In addition, the blockade of LAG-3 on Tregs abrogates their suppressive activity, indicating that this protein has an important functional role among Tregs (Anderson et al., 2016).

Another negative immune checkpoint regulator is VISTA, which is predominantly expressed on hematopoietic cells. In leukocytes, it is mostly expressed by myeloid cells (Borggrewe et al., 2018; ElTanbouly, Croteau, Noelle, & Lines, 2019). Within T lymphocytes, VISTA is upregulated in Tregs, while other T cell subsets show a lower expression of this molecule (ElTanbouly et al., 2019). VISTA is predominantly localized within endosomal compartments as it co-localizes with proteins characteristics of early and recycling endosomes (Takahashi et al., 2012; Wang, Kumar, Navarre, Casanova, & Goldenring, 2000). Its expression in TME increases with disease progression and is indicative of poor survival (Deng et al., 2019; Kuklinski et al., 2018; Liao, Zhu, Liu, & Wang, 2018; Villarroel-Espindola et al., 2018). VISTA, indeed, is able to suppress T cell responses, contributing to the immunosuppressive functions of TME (Awad, De Vlaeminck, Maebe, Goyvaerts, & Breckpot, 2018; Blando et al., 2019; Green, Wang, Noelle, & Green, 2015). In addition, its expression is highly associated with myeloid cell tumor infiltration that also correlates with poor prognosis (Awad et al., 2018).

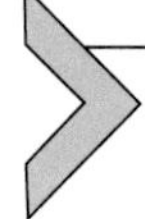

4. Immune checkpoints inhibitors (ICIs): First-line therapies to fight cancer

In the last years, ICPs have been extensively investigated as their activity can be inhibited by using blocking antibodies named immune checkpoints inhibitors (ICIs), resulting in the maintenance of a robust immune activation and an effective antitumor immune responses (Pardoll, 2012;

Stirling et al., 2022). Whereas chemotherapy and radiotherapy remain the main treatment for most cancer types, ICIs are now becoming first-line therapies for various solid and liquid tumors. The first drug employed for human therapy was ipilimumab, a monoclonal antibody against CTLA-4 used for metastatic melanoma (Bagchi, Yuan, & Engleman, 2021). Then, two anti-PD-1 antibodies were introduced, i.e., pembrolizumab and nivolumab (Jiang et al., 2019). These drugs were firstly used to treat advanced-stage melanomas, but then they have been approved for treatment of different cancer types (Liu et al., 2021). Antibodies against PD-L1 are the third class of approved ICIs, including atezolizumab, durvalumab, and avelumab (Jiang et al., 2019). Over the years, the clinical use of anti-PD-1 and anti-PD-L1 has shown greater clinical efficacy and superior tolerability compared to anti-CTLA-4 ones (Bagchi et al., 2021). Regarding promising targets, VISTA is a candidate for cancer immunotherapy too, due to its wide expression pattern and activity. The functions and expression of different markers on TILs are reported in Table 1.

ICIs were applied in clinic both as monotherapy or in combination with chemotherapy, representing a valid treatment with encouraging results

Table 1 Function and expression of the main markers that characterize TILs

Marker	Function	Expression on tils	Reference
CD45RA	T cell activation	−	Carrasco et al. (2006), Mahnke et al. (2013), and Marceaux et al. (2021)
CD45R0		+	
CCR7	Secondary lymphoid tissues homing	−	Forster et al. (2008) and Martin and Badovinac (2018)
CD27	Costimulation	−	Powell Jr. et al. (2005) and Martin and Badovinac (2018)
CD28		−	
CD103	Tissue homing	+	Stetler-Stevenson (1990), Richter and Topham (2007), Cheuk et al. (2017), and Molodtsov and Turk (2018)
CD49a		+	
CD69	T cell activation, retention, and residency within the tissue	+	Mackay et al. (2013), Baeyens et al. (2015), and Molodtsov and Turk (2018)
CXCR3	Tissue localization	+	Martin and Badovinac (2018)
CXCR6	Tissue residency	+	Muthuswamy et al. (2021)

Continued

Table 1 Function and expression of the main markers that characterize TILs—cont'd

Marker	Function	Expression on tils	Reference
HLA-DR	Activation	+	Kovacsovics-Bankowski et al. (2014) and Saraiva et al. (2021)
CD38	Lymphocyte activation, proliferation and survival	+	Kovacsovics-Bankowski et al. (2014), Malavasi et al. (2008), March et al. (2007), Ortolani et al. (1993), and Wo et al. (2019)
CD62L	Secondary lymphoid tissues homing	−	Forster et al. (2008) and Martin and Badovinac (2018)
CCR5	T cell recruitment within the TME	+	Liu et al. (2015)
CCR6		+	
Ki-67	Regulation of cell proliferation	+	Li et al. (2015) and Shirendeb et al. (2009)
IFN-γ	T cell activation, tumor cells apoptosis promotion and immune recognition	+	Gao et al. (2016), Park et al. (2017), and Burke and Young (2019)
TNF	Inflammation, anti-tumor responses, and immune system homeostasis	+	Mehta et al. (2018) and Montfort et al. (2019)
GZMB	Apoptosis induction by caspase activation	+	Cullen et al. (2010) and Li, Wang, et al. (2020)
CD25	T cell proliferation and activation	+	Ohue and Nishikawa (2019)
CD127	Interleukin 7 receptor; Th1 cell activation and Treg survival	−	Liu et al. (2006) and Ohue and Nishikawa (2019)
FoxP3	Treg generation, maintenance, and functions	+	Ohue and Nishikawa (2019) and Li, Wang, et al. (2020)
IRF4	Treg activation associated with poor survival	+	Cretney et al. (2011), Plitas et al. (2016), Li, Wang, et al. (2020), Kidani et al. (2022), and Swatler et al. (2022)
CCR8		+	
IL-10	T cell exhaustion promotion	+	Li, Wang, et al. (2020)
IL-35		+	

Table 1 Function and expression of the main markers that characterize TILs—cont'd

Marker	Function	Expression on tils	Reference
TGF-β	$CD8^+$ T cell and DC inhibition	+	
PD-1 (CD279)	Immune checkpoint protein (ICP) with suppressive function	+	Jiang et al. (2019), Ohue and Nishikawa (2019), and Liu et al. (2021)
CD39		+	Junger (2011), Antonioli et al. (2013), Cai et al. (2016), Cai, Wang, et al. (2016), Duhen et al. (2018), Canale et al. (2018), Simoni et al. (2018), Vultaggio-Poma et al. (2020), and Timperi and Barnaba (2021)
CTLA-4 (CD152)		+	Melichar et al. (2000), Rowshanravan et al. (2018), Alvisi et al. (2020), and Gibellini, De Biasi, Porta, et al. (2020)
TIM3		+	Gao et al. (2012), Kurtulus et al. (2015), Sakuishi et al. (2010), Sakuishi et al. (2013), and Yan et al. (2013)
LAG-3		+	Workman and Vignali (2003), Workman et al. (2004), Huang et al. (2004), Workman and Vignali (2005), Anderson et al. (2016), and Maruhashi et al. (2020)
TIGIT		+	Kurtulus et al. (2015), Anderson et al. (2016), and Chauvin and Zarour (2020)
VISTA		+	Green et al. (2015), Villarroel-Espindola et al. (2018), Kuklinski et al. (2018), Liao et al. (2018), Awad et al. (2018), Deng et al. (2019), and Blando et al. (2019)

(Zhou, Qiao, & Zhou, 2021). Thanks to these treatments, many patients are experiencing dramatic tumor regression. However, unfortunately not all of them respond to these therapies, and several studies are ongoing to investigate the causes of this phenomenon (De Biasi et al., 2021). In addition, some of the early responder patients develop resistance, thus experiencing tumor relapse. Therefore, a successful personalized immunotherapy is the final goal for cancer treatments (J. Liu et al., 2021).

5. Conclusion

Immuno-oncology investigates: (i) the possible reactions of immune system against cancer; (ii) cancer cells behaviour, and its responsibility in inhibiting immune responses; and (iii) relevant strategies for the manipulation of immune system to enhance anti-tumor action. Multiparametric analysis of TILs obtained by combining different technical approaches is of outstanding interest to provide a complete characterization and analysis at single-cell level that enable the identification of new molecular and cellular targets. Indeed, it is possible to study simultaneously several features of TILs, such as the adaptations to T cell phenotype and function that occur within TME and that contribute to immunosuppression. One of the most studied mechanisms is tumor-associated T cell exhaustion, that can be defined recognizing molecules present on the cell surface or in the cytoplasm. As an example, a large number of TILs present high levels of PD-1 and a lower capacity to produce IFN-γ, resulting unable to perform an effective antitumor activity (Ahmadzadeh et al., 2009; Duhen et al., 2018). These findings have important implications for modulating the immune responses in patients with cancer, and can be used for monitoring or starting therapies with the novel class of drugs such as the ICIs.

Acknowledgments

This work is supported by Associazione Italiana per la Ricerca sul Cancro, IG grant 25073 to AC. We thank Paola Paglia (ThermoFisher), Leonardo Beretta (Beckman Coulter) for their continuous support. Sara De Biasi and Lara Gibellini have been or are Marylou Ingram Scholar of the International Society for Advancement of Cytometry (ISAC) for the period 2016–2020, and 2020–2024, respectively.

References

Ahmadzadeh, M., Johnson, L. A., Heemskerk, B., Wunderlich, J. R., Dudley, M. E., White, D. E., et al. (2009). Tumor antigen-specific CD8 T cells infiltrating the tumor express high levels of PD-1 and are functionally impaired. *Blood*, *114*(8), 1537–1544. https://doi.org/10.1182/blood-2008-12-195792.

Alvisi, G., Brummelman, J., Puccio, S., Mazza, E. M., Tomada, E. P., Losurdo, A., et al. (2020). IRF4 instructs effector Treg differentiation and immune suppression in human cancer. *The Journal of Clinical Investigation*, *130*(6), 3137–3150. https://doi.org/10.1172/JCI130426.

Anderson, A. C., Joller, N., & Kuchroo, V. K. (2016). Lag-3, Tim-3, and TIGIT: Co-inhibitory receptors with specialized functions in immune regulation. *Immunity*, *44*(5), 989–1004. https://doi.org/10.1016/j.immuni.2016.05.001.

Antonioli, L., Pacher, P., Vizi, E. S., & Hasko, G. (2013). CD39 and CD73 in immunity and inflammation. *Trends in Molecular Medicine*, *19*(6), 355–367. https://doi.org/10.1016/j.molmed.2013.03.005.

Aramini, B., Masciale, V., Samarelli, A. V., Dubini, A., Gaudio, M., Stella, F., et al. (2022). Phenotypic, functional, and metabolic heterogeneity of immune cells infiltrating non–small cell lung cancer. *Frontiers in Immunology*, *13*, 959114. https://doi.org/10.3389/fimmu.2022.959114.

Awad, R. M., De Vlaeminck, Y., Maebe, J., Goyvaerts, C., & Breckpot, K. (2018). Turn back the TIMe: Targeting tumor infiltrating myeloid cells to revert cancer progression. *Frontiers in Immunology*, *9*, 1977. https://doi.org/10.3389/fimmu.2018.01977.

Baeyens, A., Fang, V., Chen, C., & Schwab, S. R. (2015). Exit strategies: S1P signaling and T cell migration. *Trends in Immunology*, *36*(12), 778–787. https://doi.org/10.1016/j.it.2015.10.005.

Bagchi, S., Yuan, R., & Engleman, E. G. (2021). Immune checkpoint inhibitors for the treatment of cancer: Clinical impact and mechanisms of response and resistance. *Annual Review of Pathology*, *16*, 223–249. https://doi.org/10.1146/annurev-pathol-042020-042741.

Baghban, R., Roshangar, L., Jahanban-Esfahlan, R., Seidi, K., Ebrahimi-Kalan, A., Jaymand, M., et al. (2020). Tumor microenvironment complexity and therapeutic implications at a glance. *Cell Communication and Signaling: CCS*, *18*(1), 59. https://doi.org/10.1186/s12964-020-0530-4.

Banerjee, H., Nieves-Rosado, H., Kulkarni, A., Murter, B., McGrath, K. V., Chandran, U. R., et al. (2021). Expression of Tim-3 drives phenotypic and functional changes in Treg cells in secondary lymphoid organs and the tumor microenvironment. *Cell Reports*, *36*(11), 109699. https://doi.org/10.1016/j.celrep.2021.109699.

Binnewies, M., Roberts, E. W., Kersten, K., Chan, V., Fearon, D. F., Merad, M., et al. (2018). Understanding the tumor immune microenvironment (TIME) for effective therapy. *Nature Medicine*, *24*(5), 541–550. https://doi.org/10.1038/s41591-018-0014-x.

Black, S., Phillips, D., Hickey, J. W., Kennedy-Darling, J., Venkataraaman, V. G., Samusik, N., et al. (2021). CODEX multiplexed tissue imaging with DNA-conjugated antibodies. *Nature Protocols*, *16*(8), 3802–3835. https://doi.org/10.1038/s41596-021-00556-8.

Blando, J., Sharma, A., Higa, M. G., Zhao, H., Vence, L., Yadav, S. S., et al. (2019). Comparison of immune infiltrates in melanoma and pancreatic cancer highlights VISTA as a potential target in pancreatic cancer. *Proceedings of the National Academy of Sciences of the United States of America*, *116*(5), 1692–1697. https://doi.org/10.1073/pnas.1811067116.

Borella, R., De Biasi, S., Paolini, A., Boraldi, F., Lo Tartaro, D., Mattioli, M., et al. (2022). Metabolic reprograming shapes neutrophil functions in severe COVID-19. *European Journal of Immunology*, *52*(3), 484–502. https://doi.org/10.1002/eji.202149481.

Borggrewe, M., Grit, C., Den Dunnen, W. F. A., Burm, S. M., Bajramovic, J. J., Noelle, R. J., et al. (2018). VISTA expression by microglia decreases during inflammation and is differentially regulated in CNS diseases. *Glia*, *66*(12), 2645–2658. https://doi.org/10.1002/glia.23517.

Bouzin, C., Brouet, A., De Vriese, J., Dewever, J., & Feron, O. (2007). Effects of vascular endothelial growth factor on the lymphocyte-endothelium interactions: Identification of

caveolin-1 and nitric oxide as control points of endothelial cell anergy. *Journal of Immunology*, *178*(3), 1505–1511. https://doi.org/10.4049/jimmunol.178.3.1505.
Brown, D. C., & Gatter, K. C. (2002). Ki67 protein: The immaculate deception? *Histopathology*, *40*(1), 2–11. https://doi.org/10.1046/j.1365-2559.2002.01343.x.
Brummelman, J., Haftmann, C., Nunez, N. G., Alvisi, G., Mazza, E. M. C., Becher, B., et al. (2019). Development, application and computational analysis of high-dimensional fluorescent antibody panels for single-cell flow cytometry. *Nature Protocols*, *14*(7), 1946–1969. https://doi.org/10.1038/s41596-019-0166-2.
Burke, J. D., & Young, H. A. (2019). IFN-gamma: A cytokine at the right time, is in the right place. *Seminars in Immunology*, *43*, 101280. https://doi.org/10.1016/j.smim.2019.05.002.
Cai, X. Y., Ni, X. C., Yi, Y., He, H. W., Wang, J. X., Fu, Y. P., et al. (2016). Overexpression of CD39 in hepatocellular carcinoma is an independent indicator of poor outcome after radical resection. *Medicine (Baltimore)*, *95*(40), e4989. https://doi.org/10.1097/MD.0000000000004989.
Cai, X. Y., Wang, X. F., Li, J., Dong, J. N., Liu, J. Q., Li, N. P., et al. (2016). High expression of CD39 in gastric cancer reduces patient outcome following radical resection. *Oncology Letters*, *12*(5), 4080–4086. https://doi.org/10.3892/ol.2016.5189.
Canale, F. P., Ramello, M. C., Nunez, N., Araujo Furlan, C. L., Bossio, S. N., Gorosito Serran, M., et al. (2018). CD39 expression defines cell exhaustion in tumor-infiltrating CD8(+) T cells. *Cancer Research*, *78*(1), 115–128. https://doi.org/10.1158/0008-5472.CAN-16-2684.
Carrasco, J., Godelaine, D., Van Pel, A., Boon, T., & van der Bruggen, P. (2006). CD45RA on human CD8 T cells is sensitive to the time elapsed since the last antigenic stimulation. *Blood*, *108*(9), 2897–2905. https://doi.org/10.1182/blood-2005-11-007237.
Chang, Q., Ornatsky, O. I., Siddiqui, I., Loboda, A., Baranov, V. I., & Hedley, D. W. (2017). Imaging Mass Cytometry. *Cytometry. Part A*, *91*(2), 160–169. https://doi.org/10.1002/cyto.a.23053.
Chattopadhyay, P. K., Winters, A. F., Lomas, W. E., 3rd, Laino, A. S., & Woods, D. M. (2019). High-parameter single-cell analysis. *Annual Review of Analytical Chemistry (Palo Alto, California)*, *12*(1), 411–430. https://doi.org/10.1146/annurev-anchem-061417-125927.
Chauvin, J. M., & Zarour, H. M. (2020). TIGIT in cancer immunotherapy. *Journal for Immunotherapy of Cancer*, *8*(2). https://doi.org/10.1136/jitc-2020-000957.
Chen, X., & Oppenheim, J. J. (2011). Resolving the identity myth: Key markers of functional CD4+FoxP3+ regulatory T cells. *International Immunopharmacology*, *11*(10), 1489–1496. https://doi.org/10.1016/j.intimp.2011.05.018.
Chen, M. L., Pittet, M. J., Gorelik, L., Flavell, R. A., Weissleder, R., von Boehmer, H., et al. (2005). Regulatory T cells suppress tumor-specific CD8 T cell cytotoxicity through TGF-beta signals in vivo. *Proceedings of the National Academy of Sciences of the United States of America*, *102*(2), 419–424. https://doi.org/10.1073/pnas.0408197102.
Cheuk, S., Schlums, H., Gallais Serezal, I., Martini, E., Chiang, S. C., Marquardt, N., et al. (2017). CD49a expression defines tissue-resident CD8(+) T cells poised for cytotoxic function in human skin. *Immunity*, *46*(2), 287–300. https://doi.org/10.1016/j.immuni.2017.01.009.
Collison, L. W., Workman, C. J., Kuo, T. T., Boyd, K., Wang, Y., Vignali, K. M., et al. (2007). The inhibitory cytokine IL-35 contributes to regulatory T-cell function. *Nature*, *450*(7169), 566–569. https://doi.org/10.1038/nature06306.
Colombo, M. P., & Piconese, S. (2007). Regulatory-T-cell inhibition versus depletion: The right choice in cancer immunotherapy. *Nature Reviews. Cancer*, 7(11), 880–887. https://doi.org/10.1038/nrc2250.

Corrales, L., Matson, V., Flood, B., Spranger, S., & Gajewski, T. F. (2017). Innate immune signaling and regulation in cancer immunotherapy. *Cell Research*, *27*(1), 96–108. https://doi.org/10.1038/cr.2016.149.

Cossarizza, A., Chang, H. D., Radbruch, A., Abrignani, S., Addo, R., Akdis, M., et al. (2021). Guidelines for the use of flow cytometry and cell sorting in immunological studies (third edition). *European Journal of Immunology*, *51*(12), 2708–3145. https://doi.org/10.1002/eji.202170126.

Cossarizza, A., Chang, H. D., Radbruch, A., Acs, A., Adam, D., Adam-Klages, S., et al. (2019). Guidelines for the use of flow cytometry and cell sorting in immunological studies (second edition). *European Journal of Immunology*, *49*(10), 1457–1973. https://doi.org/10.1002/eji.201970107.

Cossarizza, A., Chang, H. D., Radbruch, A., Akdis, M., Andra, I., Annunziato, F., et al. (2017). Guidelines for the use of flow cytometry and cell sorting in immunological studies. *European Journal of Immunology*, *47*(10), 1584–1797. https://doi.org/10.1002/eji.201646632.

Cossarizza, A., Ortolani, C., Paganelli, R., Barbieri, D., Monti, D., Sansoni, P., et al. (1996). CD45 isoforms expression on CD4+ and CD8+ T cells throughout life, from newborns to centenarians: Implications for T cell memory. *Mechanisms of Ageing and Development*, *86*(3), 173–195. https://doi.org/10.1016/0047-6374(95)01691-0.

Cretney, E., Xin, A., Shi, W., Minnich, M., Masson, F., Miasari, M., et al. (2011). The transcription factors Blimp-1 and IRF4 jointly control the differentiation and function of effector regulatory T cells. *Nature Immunology*, *12*(4), 304–311. https://doi.org/10.1038/ni.2006.

Cullen, S. P., Brunet, M., & Martin, S. J. (2010). Granzymes in cancer and immunity. *Cell Death and Differentiation*, *17*(4), 616–623. https://doi.org/10.1038/cdd.2009.206.

Dannull, J., Su, Z., Rizzieri, D., Yang, B. K., Coleman, D., Yancey, D., et al. (2005). Enhancement of vaccine-mediated antitumor immunity in cancer patients after depletion of regulatory T cells. *The Journal of Clinical Investigation*, *115*(12), 3623–3633. https://doi.org/10.1172/JCI25947.

De Biasi, S., Gibellini, L., Lo Tartaro, D., Puccio, S., Rabacchi, C., Mazza, E. M. C., et al. (2021). Circulating mucosal-associated invariant T cells identify patients responding to anti-PD-1 therapy. *Nature Communications*, *12*(1), 1669. https://doi.org/10.1038/s41467-021-21928-4.

De Biasi, S., Meschiari, M., Gibellini, L., Bellinazzi, C., Borella, R., Fidanza, L., et al. (2020). Marked T cell activation, senescence, exhaustion and skewing towards TH17 in patients with COVID-19 pneumonia. *Nature Communications*, *11*(1), 3434. https://doi.org/10.1038/s41467-020-17292-4.

De Simone, M., Arrigoni, A., Rossetti, G., Gruarin, P., Ranzani, V., Politano, C., et al. (2016). Transcriptional landscape of human tissue lymphocytes unveils uniqueness of tumor-infiltrating T regulatory cells. *Immunity*, *45*(5), 1135–1147. https://doi.org/10.1016/j.immuni.2016.10.021.

Demaria, O., Cornen, S., Daeron, M., Morel, Y., Medzhitov, R., & Vivier, E. (2019). Harnessing innate immunity in cancer therapy. *Nature*, *574*(7776), 45–56. https://doi.org/10.1038/s41586-019-1593-5.

Deng, J., Li, J., Sarde, A., Lines, J. L., Lee, Y. C., Qian, D. C., et al. (2019). Hypoxia-induced VISTA promotes the suppressive function of myeloid-derived suppressor cells in the tumor microenvironment. *Cancer Immunology Research*, *7*(7), 1079–1090. https://doi.org/10.1158/2326-6066.CIR-18-0507.

Duhen, T., Duhen, R., Montler, R., Moses, J., Moudgil, T., de Miranda, N. F., et al. (2018). Co-expression of CD39 and CD103 identifies tumor-reactive CD8 T cells in human solid tumors. *Nature Communications*, *9*(1), 2724. https://doi.org/10.1038/s41467-018-05072-0.

Dumauthioz, N., Labiano, S., & Romero, P. (2018). Tumor resident memory T cells: New players in immune surveillance and therapy. *Frontiers in Immunology*, *9*, 2076. https://doi.org/10.3389/fimmu.2018.02076.

ElTanbouly, M. A., Croteau, W., Noelle, R. J., & Lines, J. L. (2019). VISTA: A novel immunotherapy target for normalizing innate and adaptive immunity. *Seminars in Immunology*, *42*, 101308. https://doi.org/10.1016/j.smim.2019.101308.

Finn, O. J. (2017). Human tumor antigens yesterday, today, and tomorrow. *Cancer Immunology Research*, *5*(5), 347–354. https://doi.org/10.1158/2326-6066.CIR-17-0112.

Forster, R., Davalos-Misslitz, A. C., & Rot, A. (2008). CCR7 and its ligands: Balancing immunity and tolerance. *Nature Reviews. Immunology*, *8*(5), 362–371. https://doi.org/10.1038/nri2297.

Fridman, W. H., Pages, F., Sautes-Fridman, C., & Galon, J. (2012). The immune contexture in human tumours: Impact on clinical outcome. *Nature Reviews. Cancer*, *12*(4), 298–306. https://doi.org/10.1038/nrc3245.

Fritsch, R. D., Shen, X., Sims, G. P., Hathcock, K. S., Hodes, R. J., & Lipsky, P. E. (2005). Stepwise differentiation of CD4 memory T cells defined by expression of CCR7 and CD27. *Journal of Immunology*, *175*(10), 6489–6497. https://doi.org/10.4049/jimmunol.175.10.6489.

Gao, J., Shi, L. Z., Zhao, H., Chen, J., Xiong, L., He, Q., et al. (2016). Loss of IFN-gamma pathway genes in tumor cells as a mechanism of resistance to anti-CTLA-4 therapy. *Cell*, *167*(2), 397–404 e399. https://doi.org/10.1016/j.cell.2016.08.069.

Gao, X., Zhu, Y., Li, G., Huang, H., Zhang, G., Wang, F., et al. (2012). TIM-3 expression characterizes regulatory T cells in tumor tissues and is associated with lung cancer progression. *PLoS One*, *7*(2), e30676. https://doi.org/10.1371/journal.pone.0030676.

Gerner, M. Y., Kastenmuller, W., Ifrim, I., Kabat, J., & Germain, R. N. (2012). Histo-cytometry: A method for highly multiplex quantitative tissue imaging analysis applied to dendritic cell subset microanatomy in lymph nodes. *Immunity*, *37*(2), 364–376. https://doi.org/10.1016/j.immuni.2012.07.011.

Gibellini, L., De Biasi, S., Paolini, A., Borella, R., Boraldi, F., Mattioli, M., et al. (2020). Altered bioenergetics and mitochondrial dysfunction of monocytes in patients with COVID-19 pneumonia. *EMBO Molecular Medicine*, *12*(12), e13001. https://doi.org/10.15252/emmm.202013001.

Gibellini, L., De Biasi, S., Porta, C., Lo Tartaro, D., Depenni, R., Pellacani, G., et al. (2020). Single-cell approaches to profile the response to immune checkpoint inhibitors. *Frontiers in Immunology*, *11*, 490. https://doi.org/10.3389/fimmu.2020.00490.

Giesen, C., Wang, H. A., Schapiro, D., Zivanovic, N., Jacobs, A., Hattendorf, B., et al. (2014). Highly multiplexed imaging of tumor tissues with subcellular resolution by mass cytometry. *Nature Methods*, *11*(4), 417–422. https://doi.org/10.1038/nmeth.2869.

Goltsev, Y., Samusik, N., Kennedy-Darling, J., Bhate, S., Hale, M., Vazquez, G., et al. (2018). Deep profiling of mouse splenic architecture with CODEX multiplexed imaging. *Cell*, *174*(4), 968–981. e915 https://doi.org/10.1016/j.cell.2018.07.010.

Green, K. A., Wang, L., Noelle, R. J., & Green, W. R. (2015). Selective involvement of the checkpoint regulator VISTA in suppression of B-cell, but not T-cell, responsiveness by monocytic myeloid-derived suppressor cells from mice infected with an immunodeficiency-causing retrovirus. *Journal of Virology*, *89*(18), 9693–9698. https://doi.org/10.1128/JVI.00888-15.

Grossman, W. J., Verbsky, J. W., Barchet, W., Colonna, M., Atkinson, J. P., & Ley, T. J. (2004). Human T regulatory cells can use the perforin pathway to cause autologous target cell death. *Immunity*, *21*(4), 589–601. https://doi.org/10.1016/j.immuni.2004.09.002.

He, X., & Xu, C. (2020). Immune checkpoint signaling and cancer immunotherapy. *Cell Research*, *30*(8), 660–669. https://doi.org/10.1038/s41422-020-0343-4.

Hinrichs, C. S., & Rosenberg, S. A. (2014). Exploiting the curative potential of adoptive T-cell therapy for cancer. *Immunological Reviews*, *257*(1), 56–71. https://doi.org/10.1111/imr.12132.

Hintzen, R. Q., de Jong, R., Lens, S. M., & van Lier, R. A. (1994). CD27: Marker and mediator of T-cell activation? *Immunology Today*, *15*(7), 307–311. https://doi.org/10.1016/0167-5699(94)90077-9.

Huang, C. T., Workman, C. J., Flies, D., Pan, X., Marson, A. L., Zhou, G., et al. (2004). Role of LAG-3 in regulatory T cells. *Immunity*, *21*(4), 503–513. https://doi.org/10.1016/j.immuni.2004.08.010.

Jiang, Y., Chen, M., Nie, H., & Yuan, Y. (2019). PD-1 and PD-L1 in cancer immunotherapy: Clinical implications and future considerations. *Human Vaccines & Immunotherapeutics*, *15*(5), 1111–1122. https://doi.org/10.1080/21645515.2019.1571892.

Joller, N., Hafler, J. P., Brynedal, B., Kassam, N., Spoerl, S., Levin, S. D., et al. (2011). Cutting edge: TIGIT has T cell-intrinsic inhibitory functions. *Journal of Immunology*, *186*(3), 1338–1342. https://doi.org/10.4049/jimmunol.1003081.

Junger, W. G. (2011). Immune cell regulation by autocrine purinergic signalling. *Nature Reviews. Immunology*, *11*(3), 201–212. https://doi.org/10.1038/nri2938.

Keir, M. E., Butte, M. J., Freeman, G. J., & Sharpe, A. H. (2008). PD-1 and its ligands in tolerance and immunity. *Annual Review of Immunology*, *26*, 677–704. https://doi.org/10.1146/annurev.immunol.26.021607.090331.

Kennedy-Darling, J., Bhate, S. S., Hickey, J. W., Black, S., Barlow, G. L., Vazquez, G., et al. (2021). Highly multiplexed tissue imaging using repeated oligonucleotide exchange reaction. *European Journal of Immunology*, *51*(5), 1262–1277. https://doi.org/10.1002/eji.202048891.

Kidani, Y., Nogami, W., Yasumizu, Y., Kawashima, A., Tanaka, A., Sonoda, Y., et al. (2022). CCR8-targeted specific depletion of clonally expanded Treg cells in tumor tissues evokes potent tumor immunity with long-lasting memory. *Proceedings of the National Academy of Sciences of the United States of America*, *119*(7). https://doi.org/10.1073/pnas.2114282119.

Klebanoff, C. A., Gattinoni, L., & Restifo, N. P. (2006). CD8+ T-cell memory in tumor immunology and immunotherapy. *Immunological Reviews*, *211*, 214–224. https://doi.org/10.1111/j.0105-2896.2006.00391.x.

Kloppel, G., Perren, A., & Heitz, P. U. (2004). The gastroenteropancreatic neuroendocrine cell system and its tumors: The WHO classification. *Annals of the New York Academy of Sciences*, *1014*, 13–27. https://doi.org/10.1196/annals.1294.002.

Kortekaas, K. E., Santegoets, S. J., Sturm, G., Ehsan, I., van Egmond, S. L., Finotello, F., et al. (2020). CD39 identifies the CD4(+) tumor-specific T-cell population in human cancer. *Cancer Immunology Research*, *8*(10), 1311–1321. https://doi.org/10.1158/2326-6066.CIR-20-0270.

Kovacsovics-Bankowski, M., Chisholm, L., Vercellini, J., Tucker, C. G., Montler, R., Haley, D., et al. (2014). Detailed characterization of tumor infiltrating lymphocytes in two distinct human solid malignancies show phenotypic similarities. *Journal for Immunotherapy of Cancer*, *2*(1), 38. https://doi.org/10.1186/s40425-014-0038-9.

Kuklinski, L. F., Yan, S., Li, Z., Fisher, J. L., Cheng, C., Noelle, R. J., et al. (2018). VISTA expression on tumor-infiltrating inflammatory cells in primary cutaneous melanoma correlates with poor disease-specific survival. *Cancer Immunology, Immunotherapy*, *67*(7), 1113–1121. https://doi.org/10.1007/s00262-018-2169-1.

Kumar, B. V., Connors, T. J., & Farber, D. L. (2018). Human T cell development, localization, and function throughout life. *Immunity*, *48*(2), 202–213. https://doi.org/10.1016/j.immuni.2018.01.007.

Kurtulus, S., Sakuishi, K., Ngiow, S. F., Joller, N., Tan, D. J., Teng, M. W., et al. (2015). TIGIT predominantly regulates the immune response via regulatory T cells. *The Journal of Clinical Investigation*, *125*(11), 4053–4062. https://doi.org/10.1172/JCI81187.

Larmonier, N., Marron, M., Zeng, Y., Cantrell, J., Romanoski, A., Sepassi, M., et al. (2007). Tumor-derived CD4(+)CD25(+) regulatory T cell suppression of dendritic cell function involves TGF-beta and IL-10. *Cancer Immunology, Immunotherapy*, *56*(1), 48–59. https://doi.org/10.1007/s00262-006-0160-8.

Leelatian, N., Doxie, D. B., Greenplate, A. R., Mobley, B. C., Lehman, J. M., Sinnaeve, J., et al. (2017). Single cell analysis of human tissues and solid tumors with mass cytometry. *Cytometry. Part B, Clinical Cytometry*, *92*(1), 68–78. https://doi.org/10.1002/cyto.b.21481.

Li, L. T., Jiang, G., Chen, Q., & Zheng, J. N. (2015). Ki67 is a promising molecular target in the diagnosis of cancer (review). *Molecular Medicine Reports*, *11*(3), 1566–1572. https://doi.org/10.3892/mmr.2014.2914.

Li, H. L., Wang, L. H., Hu, Y. L., Feng, Y., Li, X. H., Liu, Y. F., et al. (2020). Clinical and prognostic significance of CC chemokine receptor type 8 protein expression in gastrointestinal stromal tumors. *World Journal of Gastroenterology*, *26*(31), 4656–4668. https://doi.org/10.3748/wjg.v26.i31.4656.

Liao, H., Zhu, H., Liu, S., & Wang, H. (2018). Expression of V-domain immunoglobulin suppressor of T cell activation is associated with the advanced stage and presence of lymph node metastasis in ovarian cancer. *Oncology Letters*, *16*(3), 3465–3472. https://doi.org/10.3892/ol.2018.9059.

Liechti, T., Weber, L. M., Ashhurst, T. M., Stanley, N., Prlic, M., Van Gassen, S., et al. (2021). An updated guide for the perplexed: Cytometry in the high-dimensional era. *Nature Immunology*, *22*(10), 1190–1197. https://doi.org/10.1038/s41590-021-01006-z.

Liu, J., Chen, Z., Li, Y., Zhao, W., Wu, J., & Zhang, Z. (2021). PD-1/PD-L1 checkpoint inhibitors in tumor immunotherapy. *Frontiers in Pharmacology*, *12*, 731798. https://doi.org/10.3389/fphar.2021.731798.

Liu, J. Y., Li, F., Wang, L. P., Chen, X. F., Wang, D., Cao, L., et al. (2015). CTL- vs Treg lymphocyte-attracting chemokines, CCL4 and CCL20, are strong reciprocal predictive markers for survival of patients with oesophageal squamous cell carcinoma. *British Journal of Cancer*, *113*(5), 747–755. https://doi.org/10.1038/bjc.2015.290.

Liu, W., Putnam, A. L., Xu-Yu, Z., Szot, G. L., Lee, M. R., Zhu, S., et al. (2006). CD127 expression inversely correlates with FoxP3 and suppressive function of human CD4+ T reg cells. *The Journal of Experimental Medicine*, *203*(7), 1701–1711. https://doi.org/10.1084/jem.20060772.

Luo, C. T., Liao, W., Dadi, S., Toure, A., & Li, M. O. (2016). Graded Foxo1 activity in Treg cells differentiates tumour immunity from spontaneous autoimmunity. *Nature*, *529*(7587), 532–536. https://doi.org/10.1038/nature16486.

Mackay, L. K., Rahimpour, A., Ma, J. Z., Collins, N., Stock, A. T., Hafon, M. L., et al. (2013). The developmental pathway for CD103(+)CD8+ tissue-resident memory T cells of skin. *Nature Immunology*, *14*(12), 1294–1301. https://doi.org/10.1038/ni.2744.

Maecker, H. T., & Harari, A. (2015). Immune monitoring technology primer: Flow and mass cytometry. *Journal for Immunotherapy of Cancer*, *3*, 44. https://doi.org/10.1186/s40425-015-0085-x.

Mahnke, Y. D., Brodie, T. M., Sallusto, F., Roederer, M., & Lugli, E. (2013). The who's who of T-cell differentiation: Human memory T-cell subsets. *European Journal of Immunology*, *43*(11), 2797–2809. https://doi.org/10.1002/eji.201343751.

Maibach, F., Sadozai, H., Seyed Jafari, S. M., Hunger, R. E., & Schenk, M. (2020). Tumor-infiltrating lymphocytes and their prognostic value in cutaneous melanoma. *Frontiers in Immunology*, *11*, 2105. https://doi.org/10.3389/fimmu.2020.02105.

Malavasi, F., Deaglio, S., Funaro, A., Ferrero, E., Horenstein, A. L., Ortolan, E., et al. (2008). Evolution and function of the ADP ribosyl cyclase/CD38 gene family in physiology and pathology. *Physiological Reviews*, *88*(3), 841–886. https://doi.org/10.1152/physrev.00035.2007.

Marceaux, C., Weeden, C. E., Gordon, C. L., & Asselin-Labat, M. L. (2021). Holding our breath: The promise of tissue-resident memory T cells in lung cancer. *Translational Lung Cancer Research*, *10*(6), 2819–2829. https://doi.org/10.21037/tlcr-20-819.

March, S., Graupera, M., Rosa Sarrias, M., Lozano, F., Pizcueta, P., Bosch, J., et al. (2007). Identification and functional characterization of the hepatic stellate cell CD38 cell surface molecule. *The American Journal of Pathology*, *170*(1), 176–187. https://doi.org/10.2353/ajpath.2007.051212.

Martin, M. D., & Badovinac, V. P. (2018). Defining memory CD8 T Cell. *Frontiers in Immunology*, *9*, 2692. https://doi.org/10.3389/fimmu.2018.02692.

Maruhashi, T., Sugiura, D., Okazaki, I. M., & Okazaki, T. (2020). LAG-3: From molecular functions to clinical applications. *Journal for Immunotherapy of Cancer*, *8*(2). https://doi.org/10.1136/jitc-2020-001014.

Mehta, A. K., Gracias, D. T., & Croft, M. (2018). TNF activity and T cells. *Cytokine*, *101*, 14–18. https://doi.org/10.1016/j.cyto.2016.08.003.

Melichar, B., Nash, M. A., Lenzi, R., Platsoucas, C. D., & Freedman, R. S. (2000). Expression of costimulatory molecules CD80 and CD86 and their receptors CD28, CTLA-4 on malignant ascites CD3+ tumour-infiltrating lymphocytes (TIL) from patients with ovarian and other types of peritoneal carcinomatosis. *Clinical and Experimental Immunology*, *119*(1), 19–27. https://doi.org/10.1046/j.1365-2249.2000.01105.x.

Molodtsov, A., & Turk, M. J. (2018). Tissue resident CD8 memory T cell responses in cancer and autoimmunity. *Frontiers in Immunology*, *9*, 2810. https://doi.org/10.3389/fimmu.2018.02810.

Monney, L., Sabatos, C. A., Gaglia, J. L., Ryu, A., Waldner, H., Chernova, T., et al. (2002). Th1-specific cell surface protein Tim-3 regulates macrophage activation and severity of an autoimmune disease. *Nature*, *415*(6871), 536–541. https://doi.org/10.1038/415536a.

Montfort, A., Colacios, C., Levade, T., Andrieu-Abadie, N., Meyer, N., & Segui, B. (2019). The TNF paradox in cancer progression and immunotherapy. *Frontiers in Immunology*, *10*, 1818. https://doi.org/10.3389/fimmu.2019.01818.

Mueller, S. N., & Mackay, L. K. (2016). Tissue-resident memory T cells: Local specialists in immune defence. *Nature Reviews. Immunology*, *16*(2), 79–89. https://doi.org/10.1038/nri.2015.3.

Muthuswamy, R., McGray, A. R., Battaglia, S., He, W., Miliotto, A., Eppolito, C., et al. (2021). CXCR6 by increasing retention of memory CD8(+) T cells in the ovarian tumor microenvironment promotes immunosurveillance and control of ovarian cancer. *Journal for Immunotherapy of Cancer*, *9*(10). https://doi.org/10.1136/jitc-2021-003329.

Ohue, Y., & Nishikawa, H. (2019). Regulatory T (Treg) cells in cancer: Can Treg cells be a new therapeutic target? *Cancer Science*, *110*(7), 2080–2089. https://doi.org/10.1111/cas.14069.

Okada, R., Kondo, T., Matsuki, F., Takata, H., & Takiguchi, M. (2008). Phenotypic classification of human CD4+ T cell subsets and their differentiation. *International Immunology*, *20*(9), 1189–1199. https://doi.org/10.1093/intimm/dxn075.

Ortolani, C., Forti, E., Radin, E., Cibin, R., & Cossarizza, A. (1993). Cytofluorimetric identification of two populations of double positive (CD4+,CD8+) T lymphocytes in human peripheral blood. *Biochemical and Biophysical Research Communications*, *191*(2), 601–609. https://doi.org/10.1006/bbrc.1993.1260.

Paijens, S. T., Vledder, A., de Bruyn, M., & Nijman, H. W. (2021). Tumor-infiltrating lymphocytes in the immunotherapy era. *Cellular & Molecular Immunology*, *18*(4), 842–859. https://doi.org/10.1038/s41423-020-00565-9.

Pardoll, D. M. (2012). The blockade of immune checkpoints in cancer immunotherapy. *Nature Reviews. Cancer*, *12*(4), 252–264. https://doi.org/10.1038/nrc3239.

Park, I. A., Hwang, S. H., Song, I. H., Heo, S. H., Kim, Y. A., Bang, W. S., et al. (2017). Expression of the MHC class II in triple-negative breast cancer is associated with tumor-infiltrating lymphocytes and interferon signaling. *PLoS One*, *12*(8), e0182786. https://doi.org/10.1371/journal.pone.0182786.

Plitas, G., Konopacki, C., Wu, K., Bos, P. D., Morrow, M., Putintseva, E. V., et al. (2016). Regulatory T cells exhibit distinct features in human breast cancer. *Immunity*, *45*(5), 1122–1134. https://doi.org/10.1016/j.immuni.2016.10.032.

Powell, D. J., Jr., Dudley, M. E., Robbins, P. F., & Rosenberg, S. A. (2005). Transition of late-stage effector T cells to CD27+ CD28+ tumor-reactive effector memory T cells in humans after adoptive cell transfer therapy. *Blood*, *105*(1), 241–250. https://doi.org/10.1182/blood-2004-06-2482.

Richter, M. V., & Topham, D. J. (2007). The alpha1beta1 integrin and TNF receptor II protect airway CD8+ effector T cells from apoptosis during influenza infection. *Journal of Immunology*, *179*(8), 5054–5063. https://doi.org/10.4049/jimmunol.179.8.5054.

Romero, P., Zippelius, A., Kurth, I., Pittet, M. J., Touvrey, C., Iancu, E. M., et al. (2007). Four functionally distinct populations of human effector-memory CD8+ T lymphocytes. *Journal of Immunology*, *178*(7), 4112–4119. https://doi.org/10.4049/jimmunol.178.7.4112.

Rowshanravan, B., Halliday, N., & Sansom, D. M. (2018). CTLA-4: A moving target in immunotherapy. *Blood*, *131*(1), 58–67. https://doi.org/10.1182/blood-2017-06-741033.

Roychoudhuri, R., Eil, R. L., & Restifo, N. P. (2015). The interplay of effector and regulatory T cells in cancer. *Current Opinion in Immunology*, *33*, 101–111. https://doi.org/10.1016/j.coi.2015.02.003.

Sadelain, M., Riviere, I., & Riddell, S. (2017). Therapeutic T cell engineering. *Nature*, *545*(7655), 423–431. https://doi.org/10.1038/nature22395.

Sakuishi, K., Apetoh, L., Sullivan, J. M., Blazar, B. R., Kuchroo, V. K., & Anderson, A. C. (2010). Targeting Tim-3 and PD-1 pathways to reverse T cell exhaustion and restore anti-tumor immunity. *The Journal of Experimental Medicine*, *207*(10), 2187–2194. https://doi.org/10.1084/jem.20100643.

Sakuishi, K., Ngiow, S. F., Sullivan, J. M., Teng, M. W., Kuchroo, V. K., Smyth, M. J., et al. (2013). TIM3(+)FOXP3(+) regulatory T cells are tissue-specific promoters of T-cell dysfunction in cancer. *Oncoimmunology*, *2*(4), e23849. https://doi.org/10.4161/onci.23849.

Saraiva, D. P., Azeredo-Lopes, S., Antunes, A., Salvador, R., Borralho, P., Assis, B., et al. (2021). Expression of HLA-DR in cytotoxic T lymphocytes: A validated predictive biomarker and a potential therapeutic strategy in breast cancer. *Cancers (Basel)*, *13*-(15). https://doi.org/10.3390/cancers13153841.

Schapiro, D., Jackson, H. W., Raghuraman, S., Fischer, J. R., Zanotelli, V. R. T., Schulz, D., et al. (2017). histoCAT: Analysis of cell phenotypes and interactions in multiplex image cytometry data. *Nature Methods*, *14*(9), 873–876. https://doi.org/10.1038/nmeth.4391.

Schreiber, R. D., Old, L. J., & Smyth, M. J. (2011). Cancer immunoediting: Integrating immunity's roles in cancer suppression and promotion. *Science*, *331*(6024), 1565–1570. https://doi.org/10.1126/science.1203486.

Seager, R. J., Hajal, C., Spill, F., Kamm, R. D., & Zaman, M. H. (2017). Dynamic interplay between tumour, stroma and immune system can drive or prevent tumour progression. *Convergent Science Physical Oncology*, *3*. https://doi.org/10.1088/2057-1739/aa7e86.

Shirendeb, U., Hishikawa, Y., Moriyama, S., Win, N., Thu, M. M., Mar, K. S., et al. (2009). Human papillomavirus infection and its possible correlation with p63 expression in cervical cancer in Japan, Mongolia, and Myanmar. *Acta Histochemica et Cytochemica*, *42*(6), 181–190. https://doi.org/10.1267/ahc.09030.

Simoni, Y., Becht, E., Fehlings, M., Loh, C. Y., Koo, S. L., Teng, K. W. W., et al. (2018). Bystander CD8(+) T cells are abundant and phenotypically distinct in human tumour infiltrates. *Nature*, *557*(7706), 575–579. https://doi.org/10.1038/s41586-018-0130-2.

Stanietsky, N., Simic, H., Arapovic, J., Toporik, A., Levy, O., Novik, A., et al. (2009). The interaction of TIGIT with PVR and PVRL2 inhibits human NK cell cytotoxicity. *Proceedings of the National Academy of Sciences of the United States of America*, *106*(42), 17858–17863. https://doi.org/10.1073/pnas.0903474106.

Stetler-Stevenson, W. G. (1990). Type IV collagenases in tumor invasion and metastasis. *Cancer Metastasis Reviews*, *9*(4), 289–303. https://doi.org/10.1007/BF00049520.

Stirling, E. R., Bronson, S. M., Mackert, J. D., Cook, K. L., Triozzi, P. L., & Soto-Pantoja, D. R. (2022). Metabolic implications of immune checkpoint proteins in cancer. *Cell*, *11*(1). https://doi.org/10.3390/cells11010179.

Sugasawa, H., Ichikura, T., Kinoshita, M., Ono, S., Majima, T., Tsujimoto, H., et al. (2008). Gastric cancer cells exploit CD4+ cell-derived CCL5 for their growth and prevention of CD8+ cell-involved tumor elimination. *International Journal of Cancer*, *122*(11), 2535–2541. https://doi.org/10.1002/ijc.23401.

Swatler, J., Turos-Korgul, L., Brewinska-Olchowik, M., De Biasi, S., Dudka, W., Le, B. V., et al. (2022). 4-1BBL-containing leukemic extracellular vesicles promote immunosuppressive effector regulatory T cells. *Blood Advances*, *6*(6), 1879–1894. https://doi.org/10.1182/bloodadvances.2021006195.

Takahashi, S., Kubo, K., Waguri, S., Yabashi, A., Shin, H. W., Katoh, Y., et al. (2012). Rab11 regulates exocytosis of recycling vesicles at the plasma membrane. *Journal of Cell Science*, *125*(Pt 17), 4049–4057. https://doi.org/10.1242/jcs.102913.

Tan, Y. S., & Lei, Y. L. (2019). Isolation of tumor-infiltrating lymphocytes by Ficoll-Paque density gradient centrifugation. *Methods in Molecular Biology*, *1960*, 93–99. https://doi.org/10.1007/978-1-4939-9167-9_8.

Tanaka, A., & Sakaguchi, S. (2017). Regulatory T cells in cancer immunotherapy. *Cell Research*, *27*(1), 109–118. https://doi.org/10.1038/cr.2016.151.

Timperi, E., & Barnaba, V. (2021). CD39 Regulation and Functions in T Cells. *International Journal of Molecular Sciences*, *22*(15). https://doi.org/10.3390/ijms22158068.

Triebel, F., Jitsukawa, S., Baixeras, E., Roman-Roman, S., Genevee, C., Viegas-Pequignot, E., & Hercend, T. (1990). LAG-3, a novel lymphocyte activation gene closely related to CD4. *Journal of Experimental Medicine*, *171*(5), 1393–1405. https://doi.org/10.1084/jem.171.5.1393.

Tsakiroglou, A. M., Fergie, M., Oguejiofor, K., Linton, K., Thomson, D., Stern, P. L., et al. (2020). Spatial proximity between T and PD-L1 expressing cells as a prognostic biomarker for oropharyngeal squamous cell carcinoma. *British Journal of Cancer*, *122*(4), 539–544. https://doi.org/10.1038/s41416-019-0634-z.

Villarroel-Espindola, F., Yu, X., Datar, I., Mani, N., Sanmamed, M., Velcheti, V., et al. (2018). Spatially resolved and quantitative analysis of VISTA/PD-1H as a novel immunotherapy target in human non-small cell lung cancer. *Clinical Cancer Research*, *24*(7), 1562–1573. https://doi.org/10.1158/1078-0432.CCR-17-2542.

Vultaggio-Poma, V., Sarti, A. C., & Di Virgilio, F. (2020). Extracellular ATP: A feasible target for cancer therapy. *Cell*, *9*(11). https://doi.org/10.3390/cells9112496.

Wagner, J., Rapsomaniki, M. A., Chevrier, S., Anzeneder, T., Langwieder, C., Dykgers, A., et al. (2019). A single-cell atlas of the tumor and immune ecosystem of human breast cancer. *Cell*, *177*(5), 1330–1345.e18. https://doi.org/10.1016/j.cell.2019.03.005.

Wang, X., Kumar, R., Navarre, J., Casanova, J. E., & Goldenring, J. R. (2000). Regulation of vesicle trafficking in madin-darby canine kidney cells by Rab11a and Rab25. *The*

Journal of Biological Chemistry, 275(37), 29138–29146. https://doi.org/10.1074/jbc.M004410200.

Wang, Y., Zong, B., Yu, Y., Wang, Y., Tang, Z., Chen, R., et al. (2021). Ki67 index changes and tumor-infiltrating lymphocyte levels impact the prognosis of triple-negative breast cancer patients with residual disease after neoadjuvant chemotherapy. *Frontiers in Oncology, 11*, 668610. https://doi.org/10.3389/fonc.2021.668610.

Wei, W. Z., Jacob, J. B., Zielinski, J. F., Flynn, J. C., Shim, K. D., Alsharabi, G., et al. (2005). Concurrent induction of antitumor immunity and autoimmune thyroiditis in CD4+ CD25+ regulatory T cell-depleted mice. *Cancer Research, 65*(18), 8471–8478. https://doi.org/10.1158/0008-5472.CAN-05-0934.

Wing, K., Onishi, Y., Prieto-Martin, P., Yamaguchi, T., Miyara, M., Fehervari, Z., et al. (2008). CTLA-4 control over Foxp3+ regulatory T cell function. *Science, 322*(5899), 271–275. https://doi.org/10.1126/science.1160062.

Wo, Y. J., Gan, A. S. P., Lim, X., Tay, I. S. Y., Lim, S., Lim, J. C. T., et al. (2019). The roles of CD38 and CD157 in the solid tumor microenvironment and cancer immunotherapy. *Cell, 9*(1). https://doi.org/10.3390/cells9010026.

Woo, S. R., Corrales, L., & Gajewski, T. F. (2015). Innate immune recognition of cancer. *Annual Review of Immunology, 33*, 445–474. https://doi.org/10.1146/annurev-immunol-032414-112043.

Workman, C. J., Cauley, L. S., Kim, I. J., Blackman, M. A., Woodland, D. L., & Vignali, D. A. (2004). Lymphocyte activation gene-3 (CD223) regulates the size of the expanding T cell population following antigen activation in vivo. *Journal of Immunology, 172*(9), 5450–5455. https://doi.org/10.4049/jimmunol.172.9.5450.

Workman, C. J., & Vignali, D. A. (2003). The CD4-related molecule, LAG-3 (CD223), regulates the expansion of activated T cells. *European Journal of Immunology, 33*(4), 970–979. https://doi.org/10.1002/eji.200323382.

Workman, C. J., & Vignali, D. A. (2005). Negative regulation of T cell homeostasis by lymphocyte activation gene-3 (CD223). *Journal of Immunology, 174*(2), 688–695. https://doi.org/10.4049/jimmunol.174.2.688.

Yan, J., Zhang, Y., Zhang, J. P., Liang, J., Li, L., & Zheng, L. (2013). Tim-3 expression defines regulatory T cells in human tumors. *PLoS One, 8*(3), e58006. https://doi.org/10.1371/journal.pone.0058006.

Zanini, G., De Gaetano, A., Selleri, V., Savino, G., Cossarizza, A., Pinti, M., et al. (2021). Mitochondrial DNA and exercise: Implications for health and injuries in sports. *Cell, 10*(10), 2575. https://doi.org/10.3390/cells10102575.

Zhang, Y., Guan, X. Y., & Jiang, P. (2020). Cytokine and chemokine signals of T-cell exclusion in tumors. *Frontiers in Immunology, 11*, 594609. https://doi.org/10.3389/fimmu.2020.594609.

Zhou, F., Qiao, M., & Zhou, C. (2021). The cutting-edge progress of immune-checkpoint blockade in lung cancer. *Cellular & Molecular Immunology, 18*(2), 279–293. https://doi.org/10.1038/s41423-020-00577-5.

CHAPTER FOUR

Assessing chromosomal abnormalities in leukemias by imaging flow cytometry

Stephanie J. Lam[a,b], Henry Y.L. Hui[c], Kathy A. Fuller[c], and Wendy N. Erber[b,c,*]

[a]Department of Haematology, Fiona Stanley Hospital, Murdoch, WA, Australia
[b]Department of Haematology, PathWest Laboratory Medicine, Nedlands, WA, Australia
[c]School of Biomedical Sciences, The University of Western Australia, Perth, WA, Australia
*Corresponding author: e-mail address: wendy.erber@uwa.edu.au

Contents

Abstract

Chromosome analysis assists in the diagnostic classification and prognostication of leukemias. It is typically performed by karyotyping or fluorescent in situ hybridization (FISH) on glass slides. Flow cytometry offers an alternative high throughput automated methodology to analyze chromosomal content. With the advent of imaging flow cytometers, specific chromosomes and regions of interest can be identified and enumerated within specific cell types. The inclusion of immunophenotyping increases the specificity of this technique to ensure only the leukemic cell is analyzed. With many thousands of cells acquired, and neoplastic cells of interest identified by antigen expression, this technology has expanded the role of flow cytometry for cytogenomics in oncology. Applications to date have focused on hematological malignancies to detect aneuploidy (chromosome gains and losses) and structural defects (e.g., deletions;

Methods in Cell Biology, Volume 195
ISSN 0091-679X
https://doi.org/10.1016/bs.mcb.2023.04.001

translocations) of diagnostic or prognostic significance at the time of diagnosis. With limits of detection of 1 cytogenetically abnormal cell in 100,000, also makes this new flow cytometry protocol eminently suitable for monitoring low level disease, detecting clonal evolution after therapy and identifying circulating tumor cells. The technique is equally applicable to solid tumors, many of which have chromosomal aberrations, with selection of appropriate immunophenotypic markers and FISH probes.

1. Background

Hematologic malignancies account for approximately 10% of cancers in the developed world, including leukemia, lymphoma and myeloma. Traditionally, the diagnostic work-up of these neoplasms involves integrating the information from several laboratory tests including morphology by light microscopy, phenotyping by flow cytometry or immunohistochemistry, genetics by chromosome analysis (karyotype and fluorescence in situ hybridization (FISH)) and histology. This holistic assessment gives a complete diagnostic picture and WHO classification, indicates likely prognosis, identifies potential markers to be used for disease monitoring and can guide therapy.

Flow cytometry is one of the most important diagnostic techniques with its prime roles being to establish the cell lineage, stage of differentiation and disease-association (and thereby diagnosis and classification) by virtue of antigens expressed by the neoplastic cells. For example, in chronic lymphocytic leukemia (CLL), leukemic cells have a characteristic B-lineage phenotype with aberrant CD5 expression, often in conjunction with CD23 expression. The applications extend to prognostic predictors, potential therapeutic targets and residual disease post-therapy. Testing can be performed rapidly on blood or freshly aspirated bone marrow, is multiparametric enabling multiple antigens to be assessed simultaneously and can detect both surface and intracellular (cytoplasmic and nuclear) antigens. However, the cells are not directly visualized and so it can only be used to determine the number of cells expressing an antigen and the intensity of expression (i.e., mean fluorescence intensity).

The antigenic profile identified by flow cytometry can also indicate the presence of some chromosomal defects as some immunophenotypic patterns have specific genetic associations. As such, these antigenic alterations can be used as surrogate markers of underlying genetic defects. Examples include in B-lymphoblastic leukemia/lymphoma where lack of expression of CD10 and TdT are associated with 11q23 abnormalities involving the *KMT2A* gene; and, intense expression of CD10 with dim CD9 and CD20 with t(12;21)(p13;q22); *ETV6-RUNX1*. Another example is multiple myeloma where small plasma cell morphology and expression of CD20 are associated with t(11;14)(q13;q32); *IGH-CCND1*. Further, in acute myeloid leukemia (AML) with mutated *NPM1*, the blast cells are generally CD34-negative, and, in acute promyelocytic leukemia (APML) with the *PML-RARA* genetic alteration, there is characteristically bright CD33 expression, lack of expression of CD15, CD34 and HLA-DR antigens and CD2 is commonly expressed. Since chromosomal aberrations make up part of the classification paradigm for hematological malignancies, these phenotype-genotype associations can be a useful screen for chromosomal defects.

Flow cytometry has now progressed to enable phenotype and chromosomal analysis to be integrated into one test with direct visualization of both the antigen and chromosome. This has been achieved using the capability of imaging flow cytometers. These instruments, first introduced in the late 1970s (Cambier, Kay, & Wheeless, 1979; Kachel, Benker, Lichtnau, Valet, & Glossner, 1979), are similar to traditional flow cytometers with the addition of optical microscope objectives and digital cameras. With this additional capability, imaging flow cytometers capture and record high power digital images of each cell, in addition to standard flow cytometric quantitative numerical data (Barteneva, Fasler-Kan, & Vorobjev, 2012; Basiji, Ortyn, Liang, Venkatachalam, & Morrissey, 2007). As such, cell morphology (size, shape) and localized fluorescent signals (i.e., membrane, cytoplasm, nucleus) and intensity can all be visualized and measured.

This additional functionality offered by imaging flow cytometers has been used in leukemia diagnostics to assess subtypes of AML, and specifically where there is altered localization of a protein. First, in APML the abnormal pattern of PML bodies resulting from the t(15;17); *PML-RARA* can be seen with imaging flow cytometry as a diffuse pattern of staining, which differs from the normal clumped pattern (Grimwade, Fuller, & Erber, 2017; Grimwade, Gudgin, Bloxham, Scott, & Erber, 2011). Another example is AML with mutated *NPM1* where mutations in the nucleophosmin 1 gene cause translocation of the NPM protein from the nucleus to the cytoplasm. This translocated cytoplasmic NPM can be seen with visual imagery with imaging flow cytometry in AML NPM+ cases, whereas in other types of AML the NPM is within the nucleus (Grimwade et al., 2017, 2012). These two examples illustrate the additional value of imaging flow cytometry for detecting surrogate markers of chromosomal changes in leukemia; in both examples, the altered protein pattern and localization cannot be identified by standard flow cytometry.

Another feature of imaging flow cytometers is "extended depth of field" (EDF) capability. EDF means that objects as small as 2 μm, and in different focal planes within approximately 15 μm, are in precise focus. This results in a high-resolution image of the entire cell but with individual cellular components being in focus simultaneously (Basiji et al., 2007). The EDF functionality has led to imaging flow cytometers being able to be used to identify small intracellular "objects." This has been applied to detect chromosomes by FISH. This work, pioneered by Minderman et al., developed FISH in suspension (FISH-IS) to detect trisomy 8 in AML (Minderman et al., 2012). Through the acquisition of 10,000 events they demonstrated that FISH-IS could detect aneuploidy in AML and was more sensitive than conventional FISH techniques (which classically only assess 200 nuclei).

FISH of cells in suspension progressed to include immunophenotypic antigen expression simultaneously with chromosomal analysis (Fuller, Bennett, Hui, Chakera, & Erber, 2016; Hui et al., 2018). The addition of immunophenotyping increased the precision of testing as each cell is first identified by its antigenic profile and then assessed for specific chromosomal changes. It is still based on standard flow cytometry principles and integrates objective visualization of the cell, immunophenotype and nuclear FISH probe signals together with quantitative numerical data. With the acronym "immuno-flowFISH," the methodology entails antibody labeling, DNA denaturation followed by FISH probe hybridization and acquisition of cells using an imaging flow cytometer (e.g., the Amnis ImageStream®X Mark II). With appropriate software (e.g., IDEAS) cells are analyzed for the antigen profile and then interrogated for the chromosome/s of interest. Immuno-flowFISH has been applied to assess ploidy (e.g., monosomy; trisomy), chromosomal translocations and loss of chromosomal regions (e.g., del(17p)) in hematological malignancies (Erber et al., 2020; Hui et al., 2019, 2018; Hui, Stanley, Clarke, Erber, & Fuller, 2021).

The specific attributes of immuno-flowFISH include:

1. No prior cell separation is required
2. Many thousands of whole cells are acquired and assessed, and not just the nucleus (as with standard slide-based FISH)
3. Sensitivity to detect a leukemic cell with a chromosomal abnormality in 10^5 cells
4. High specificity as only the cell of interest, based on antigen expression, is assessed for chromosomal alterations
5. Suitability for the assessment of ploidy and structural chromosomal changes

6. Digital images can be assessed and downloaded

Here, we provide details of the immuno-flowFISH protocol illustrated with examples of CLL and plasma cell myeloma.

2. Advantages

The immuno-flowFISH methodology described has been utilized to identify both numeric and structural chromosomal abnormalities in both acute and chronic hematological malignancies. The power of this technique comes from the incorporation of immunophenotyping which ensures that the FISH signals are only assessed in the cell type of interest based on antigen expression. This is facilitated by being able to see the cells and their chromosomal signal. The addition of a nuclear counterstain and visual images verifies the specificity of the FISH signals. Central to the procedure are:

1. There is no need for any pre-analytical cell isolation or sorting
2. Whole cells are analyzed in suspension (analogous to flow cytometry)
3. The power of the digital images which allows direct visualization of cells, their nucleus, cell phenotype and chromosomal signal simultaneously
4. The extended depth of field enabling the cell and FISH spot count images to be in focus
5. High specificity through the positive identification of cells of interest based on immunophenotype
6. The high number of cells that can be analyzed increasing sensitivity

2.1 Technical considerations

The protocol is described in detail and illustrated in Fig. 1. It requires key procedures to be followed in an orderly sequence from specific sample preparation and probe hybridization to sample acquisition in the imaging flow cytometer. The procedure has been developed by incorporation and

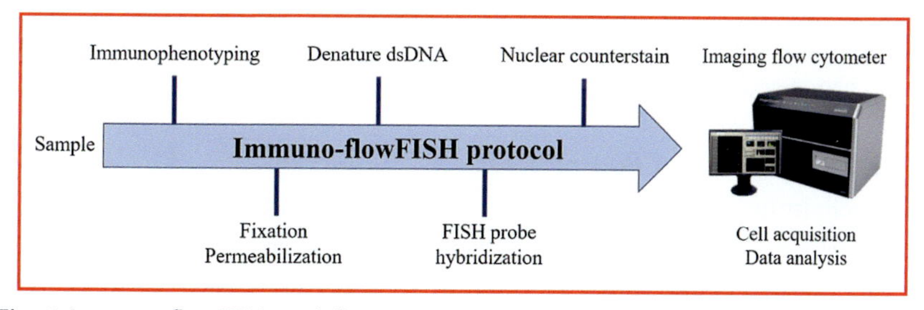

Fig. 1 Immuno-flowFISH work flow.

modifications being made to traditional multiparameter flow cytometry and slide-based FISH methods. This required a number of key components to be considered to ensure:

1. The cells are intact with preserved membranes
2. Cellular antigens are not destroyed by the need for DNA denaturation required for FISH probe hybridization
3. The need to permeabilize cell membranes to enable FISH probes to access the intranuclear DNA
4. Appropriate antibody and FISH probe fluorophores to be used that are detectable by digital imagery
5. Compensation and spectral overlap issues to be optimized to maximize FISH probe detection for enumeration
6. Data analytical methods that incorporate visual images and quantitative numerical data

The cells in suspension are first incubated with antibody, post-fixed and the membrane permeabilized prior to preparation of the DNA for probe hybridization. Preparation time is <36 h, faster than traditional slide-based FISH, and the analysis is largely automated. An explanation for these technical aspects are discussed.

2.1.1 Sample types and cells

The sample types that can be studied are those in which viable cells are in cell suspension. To date, this cytogenomic protocol has been performed on fresh blood and bone marrow samples as well as cryopreserved biobanked samples and cell lines to analyze hematological malignancies. Just as for standard flow cytometry, the method could also be performed on single cell suspensions prepared from cells extracted from tissue biopsies. Red cell lysis is generally performed prior to antibody labeling. There is no need for cell isolation or malignant cell purification prior to analysis; however, mononuclear cell isolation can be used. Samples should be in EDTA anticoagulant and processing commenced within 24 h of collection to maximize cell integrity. All specimens can be maintained at room temperature (18–22 °C) if they are to be processed immediately.

2.1.2 Antibodies and fluorophores

The cell immunophenotyping with fluorescent-conjugated antibodies is performed prior to FISH probe hybridization; reversing the order results in high non-specific antibody binding to the cells (Fuller et al., 2016) (Fig. 1). Antibody selection, combined with fluorophore chemistry, is the most critical aspect of reliable immunophenotyping. Antibodies commonly used are generally those used in standard diagnostic flow cytometry and are selected to best identify the cell of interest (i.e., leukemic cell) while still having an appropriate negative control (e.g., CD3 T-lymphocytes) in the test. Each clinical scenario determines the most appropriate antibodies to be used; for example, CD5 and CD19 have been used to identify CLL; CD38 and CD138 for plasma cells (Erber et al., 2020; Hui et al., 2019). Each antibody needs to be worked up individually to ensure binding is maintained during the course of processing. The number of antibodies that can be included in a test depends on the clinical indication, the known malignant cell phenotype, the fluorophore repertoire available and laser configuration of the instrument.

A variety of fluorophores are available with unique excitation and emission characteristics depending upon the types of lasers used. The most appropriate fluorophore must be selected for each antibody to maximize antigen detection, as per standard flow cytometry, but taking into account the need for the fluorophore to be unaffected by heat and acid and high laser powers. Further, the FISH probes are also fluorophore-conjugated and the spectral characteristics of these must also be considered. Conventional organic- or protein-based fluorescent molecules used in standard multiparameter flow cytometry (such as FITC, APC, PE) and tandem combinations of these fluorescent molecules (e.g., APCCy7 and PECy7) lose their fluorescence intensity after heat and acid conditions used for DNA denaturation processing and makes them unsuitable for immuno-flowFISH. The newer synthetic polymers or chemically-modified fluorophores (e.g., Brilliant and Alexa Fluor) have greater brightness, photo-stability, resistance to extreme pH and temperature changes and have a more specific or discrete emission spectrum. Of note, other than polymers, most tandem dyes are unsuitable for the acid and high temperature conditions required. Fluorophores such as BB515, BV480 and AF647 give the best fluorescence intensity and are recommended as antibody conjugates (Fuller et al., 2016; Hui et al., 2018).

To preserve the antibody-fluorophore conjugate, BS3, a membrane-impermeable, non-denaturing *N*-hydroxysuccinimide (NHS) ester is added to cross-link membrane proteins (Fuller et al., 2016). This stabilizes the antibody-antigen complex on the cell membrane prior to commencing the FISH process. A post-phenotyping fixation step with 4% formaldehyde preserves cellular integrity whilst retaining sufficient immunophenotyping fluorescent signal for detection. Alternate fixatives, such as Carnoy's solution, are less suitable as they caused shedding of some antibodies and/or quenching of the fluorophore signal (Fuller et al., 2016).

2.1.3 FISH probes and fluorophores

As detailed in the protocol, FISH probe hybridization is performed after antibody labeling, cross-linking and fixation (Fig. 1). This requires initial denaturation of double-stranded DNA followed by membrane permeabilization with a non-ionic detergent to facilitate probe entry. Overnight hybridization (minimum 24 h) maximizes probe binding to complementary DNA sequences. Both numeric and locus specific probes, generally of 300–600 kb, have been used. Most commercially available FISH probes have been designed for fluorescent microscopy and are generally conjugated with fluorophores such as FITC and TexasRed. Fluorophores of choice are Spectrum Orange (SO), Spectrum Green (SG), and Carboxytetramethylrhodamine (TAMRA) due to the clarity and brightness of their signals captured by imaging flow cytometry, allowing clear visualization of the hybridized probe against background staining (Erber et al., 2020; Fuller, Hui, Stanley, & Erber, 2021; Hui et al., 2019, 2018).

2.1.4 Instrumentation and analysis

In imaging flow cytometers the cells in fluid suspension pass in single-file through finely focused laser beams. A range of excitation lasers (e.g., 405, 488, 561, 592, 642 nm) may be required depending on the fluorophores used for both the antibody and FISH probe. The best images are obtained with the ×60 objective using the EDF capability. Imaging flow cytometry devices have lower throughput than standard flow cytometers. For example, the Amnis ImageStream®X Mark II has a throughput speed of up to 6000 cells/s; but with chromosomal analysis acquisition speeds of <1000 cells/s are required to achieve the highest sensitivity image analysis.

There are many qualitative and quantitative analytical tools that can be used to analyze leukemic cells for chromosomal assessment. These include:

1. Morphological and phenotypic identification of individual cells
2. "Features" and "masks" to ensure nuclear localization of FISH signals
3. FISH "spot count" enumeration
4. Spot count ratios comparing normal and neoplastic cells

Single cells are first identified in a scatter plot of the "Aspect Ratio" vs "Brightfield Area" (Ch01) (Fig. 2). The antigen expression is then used to identify the cells of interest. Cells that are in focus, single and non-dividing cells (based on nuclear marker intensity) are then assessed for FISH probe "spots." Cells with hybridized FISH probe signals are identified by "Max

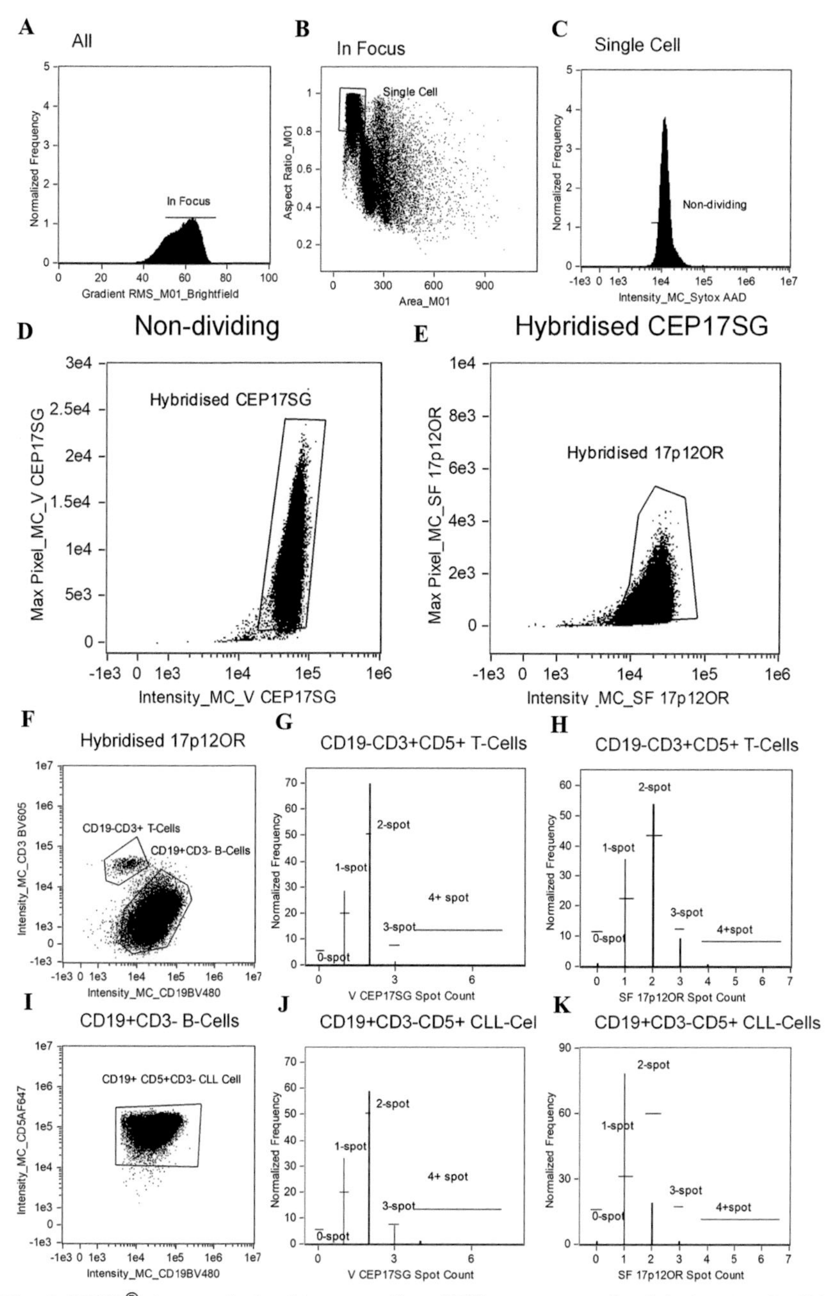

Fig. 2 IDEAS® data analysis of immuno-flow FISH assessment for deletion 17p in CLL. In focus images (A) are selected for analysis. Single (B) nucleated non-dividing (C) cells with CEP17 (D) and 17p12 (E) hybridization signals are selected. $CD19^{+}CD3^{-}CD5^{+}$ CLL cells are gated (F) and the spot counts for CEP17 (G) and 17p12 (H) FISH probes calculated. $CD19^{-}CD3^{+}CD5^{+}$ T-lymphocytes are gated (I) and the spot counts for CEP17 (J) and 17p12 (K) FISH probes calculated.

pixel" vs "Intensity" scatterplots for each probe (Ch02/Ch03/Ch04). The "Spot Count" feature is used to count the number of FISH probe "spots" per cell overlying the counterstained nucleus. This calculation analyses the pixel intensity of the probe spots to cell background ratio, the radius of the spots as well as the intensity of the probe fluorescence. Overlapping spots, due to alignment with the laser or close physical localization within the cell, will give double the fluorescence intensity of a single FISH signal. The spot count profile must therefore be corrected to account for these events (as described by Minderman et al., 2012). In addition, non-specific FISH spots that are external to the nucleus are excluded from analyses.

Normal (diploid) cells will have a mean spot count of close to two for each probe. FISH spot count data of the leukemic cell can then be compared with the normal control cells. In neoplastic cells, there may be aneuploidy (e.g., monosomy, trisomy), deletions of a region giving a single FISH signal, or, fused signals resulting from chromosomal translocations. The data is presented as the number and percent of all cells, and those with the phenotype of interest, with the specific chromosomal change. Data can be presented in the same fashion as standard FISH, i.e., 2R2G (normal), 3G (trisomy), 1R2G (deletion), 1R1G2F (for translocations), where "R" and "G" refer to the color of the probe, "red" and "green," respectively. Loss of chromosomal material (e.g., monosomy or deletions) will give a mean spot count of <2, and a ratio of <1 when compared with the controls. Cells with additional chromosomes or regions (e.g., trisomy) will have a mean spot count of >2 and a ratio to normal cells of >1. Generating a spot count ratio provides a useful "screen" for chromosomal gains and losses; this numerical data can then be verified by digital images (Figs. 3 and 4).

Large numbers of cells can be acquired and analyzed (up to 500,000 per analysis). This leads to high sensitivity and specificity of the protocol. To date, the lowest reported limit of detection of a chromosomal defect by immuno-flowFISH is when present in <0.1% of cells in a sample (Erber et al., 2020; Fuller et al., 2021; Hui et al., 2019). Further refinements may further improve this level for detecting rare events and applicability in minimal residual disease assessment.

3. Clinical applications

The value of imaging flow cytometry to detect numerical and structural chromosomal abnormalities in hematological malignancies will now be described, citing examples of chronic lymphocytic leukemia and plasma cell myeloma.

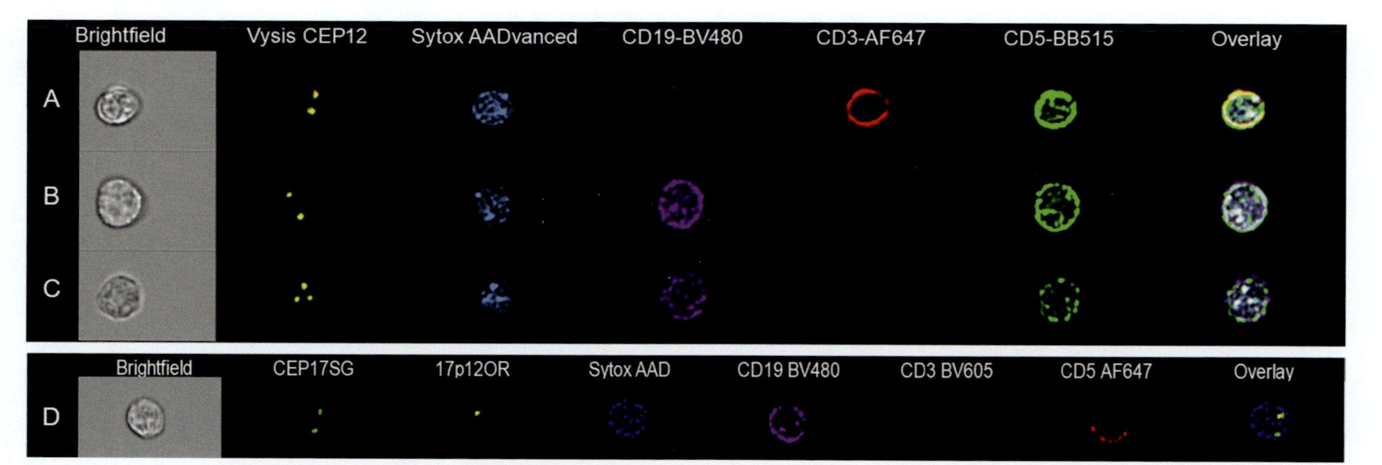

Fig. 3 Immuno-flowFISH image galleries of chronic lymphocytic leukemia. (A) Normal T-lymphocyte (CD3-AF647$^+$/CD5-BB515$^+$) with normal 2 CEP12 signals (Vysis CEP12). (B) CLL cell (CD19-BV480$^+$/CD5-BB515$^+$) with two normal CEP12 signals (Vysis CEP12). (C) CLL cell with trisomy 12 (CD19-BV480$^+$/CD5-BB515$^+$ cell and 3 CEP12 signals) (Vysis CEP12). (D) CLL cell with del(17p). There is 1 spot for 17p12 (17p12OR) indicating a deletion from one chromosome 17. The presence of 2 CEP17 signals (CEP17SG) indicates two 17 centromeres.

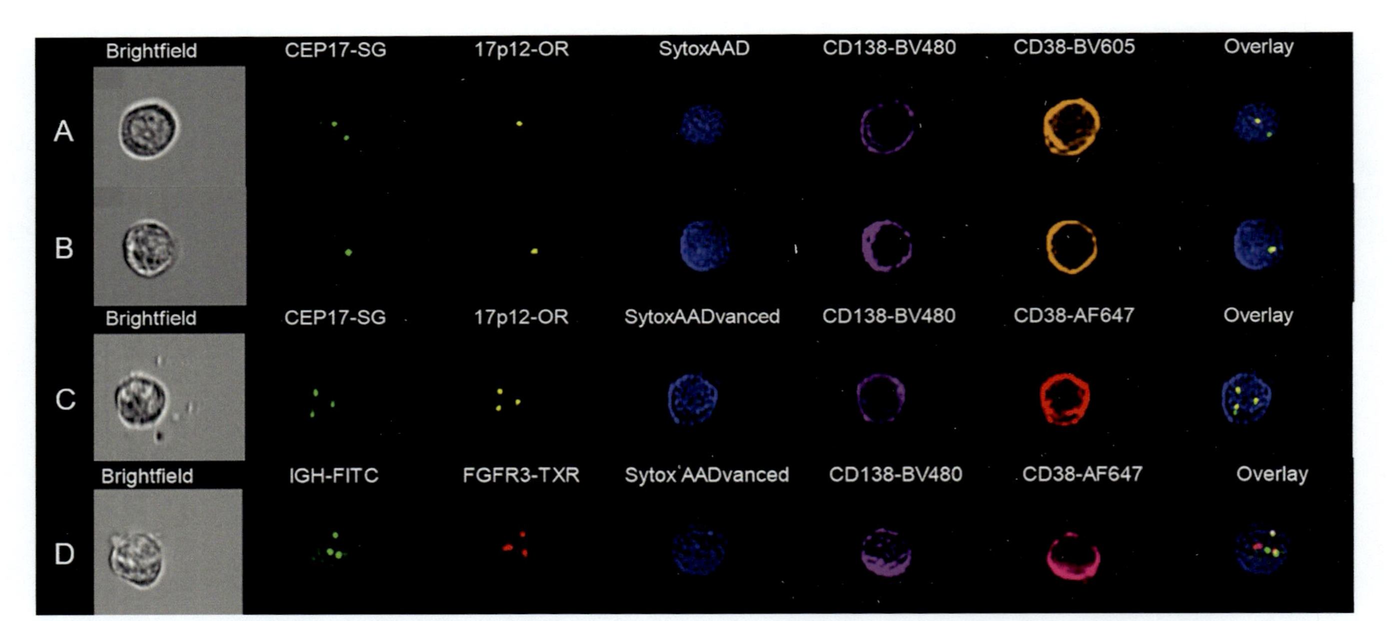

Fig. 4 Immuno-flowFISH image galleries of plasma cells in multiple myeloma. (A) Deletion 17p (1 FISH signal for 17p12 and 2 for CEP17) in a plasma cell (CD138-BV480^{+}/CD38-BV605^{+}). (B) Monosomy 17 evidenced by one FISH signal for both CEP17 and 17p12 in a plasma cell (CD138-BV480^{+}/CD38-BV605^{+}). (C) Trisomy 17 in a plasma cell (CD138-BV480^{+}/CD38-AF647^{+}) with three signals for both CEP17 and 17p12 probes. (D) Plasma cell with t(4;14); *IGH-FGFR3* evidenced by the 1G1R2F FISH pattern in the overlay image. There are three IGH and three FGFR3 signals: two fused (2F) signals from the reciprocal chromosomal translocation and one IGH (1G) and one FGFR3 (1R) from the non-involved chromosomes 14 and 4, respectively.

3.1 Chronic lymphocytic leukemia

Chronic lymphocytic leukemia (CLL) is the commonest hematological cancer in the Western world. It is a genetically heterogeneous disease with treatment and prognosis varying based on chromosomal abnormalities. As treatments advance and more targeted therapies become available, clinicians have identified that certain therapies are more effective at treating patients with specific cytogenetic changes. Chromosome abnormalities are detectable in up to 80% of patients on traditional cytogenetic analysis performed by FISH on interphase cells. Of these deletions of 11q, 13q and 17p and trisomy 12, are the most common and of prognostic significance. Trisomy 12 occurs in 10–20% of patients, has an intermediate prognosis, and is associated with increased risk of progression to the aggressive Richter's transformation. Deletion of 17p13 (or "del(17p)") occurs in up to 10% of patients at diagnosis, and 30% at relapse following treatment. It is the highest risk prognostic category, with the shortest progression-free and overall survival (Scarfo, Ferreri, & Ghia, 2016) Further, these patients do not have a durable response to standard chemotherapy and, to improve survival, require alternate targeted inhibitor approaches. Minimal residual disease (MRD) negativity has been shown to be prognostically significant with patients who achieve this milestone, remaining disease free for prolonged periods (Thompson et al., 2016). Presently, MRD assays are performed using flow cytometry or molecular based assessment to a detection threshold of 10^{-4} with conventional FISH testing unsuitable due to the small number of cells assessed.

The immuno-flowFISH descriptions for CLL have focused on detection of trisomy 12 and del(17p) (Erber et al., 2020; Fuller et al., 2021; Hui et al., 2019, 2018). The antibody panels used (i.e., CD3, CD5 and CD19 conjugated with BV480, BV605 and AF647 fluorophores) have been sufficient to differentiate CLL (CD5, CD19 co-expression) from T-lymphocytes (CD3, CD5-dual positive). Further antibodies could be assessed for inclusion to address different questions (e.g., CD38, ZAP70 expression, cytogenetics and prognosis). The probes, assessing the centromere of chromosome 12 and 17p12 region were to assess the prognostically important trisomy 12 and del(17p) (Fig. 3). A CEP17 centromeric probe was included as a control used in the assessment for possible del(17p). In these studies, the CD3-positive population was a normal cellular control which can be used to confirm diploid signals for each probe, and to generate a ratio of FISH "spot counts" between normal (T) and neoplastic (CLL) cells. A ratio of CLL/T spot ratio of >1 indicates additional chromosomal material (i.e., hyperploidy) whereas a ratio of <1 hypoploidy.

The lowest limit of detection for a FISH-detectable abnormality in CLL cells has been reported to be 10^{-5} (Fuller et al., 2021). This high sensitivity and specificity exceeds current clinical practice guidelines (10^{-4}), and is achievable due to the many thousands of cells assessed and positive phenotypic identification of the CLL cells. This ultra-sensitive high precision immuno-flowFISH method therefore adds a new dimension to chromosomal analysis in CLL, both at diagnosis and potential to be used for MRD monitoring. Multi-FISH probing is also possible utilizing CD3, CD5, CD19 antibodies and incorporating the CEP12, CEP17 and 17p12 FISH probes in one analysis (Hui et al., 2021).

3.2 Plasma cell myeloma

Plasma cell myeloma or "multiple myeloma" is characterized by monoclonal plasma cells, predominantly in the bone marrow. It is the second most common blood cancer and accounts for 10–15% of hematological neoplasms. At diagnosis, approximately 40% of patients will have cytogenetic abnormalities detected on FISH studies on a bone marrow sample. A proportion of diagnostic bone marrow samples do not yield a FISH result due to low plasma cell numbers in the aspirate with plasma cell selection or isolation methods required or due to other technical difficulties. Further, most laboratories score only 100–200 plasma cells per sample and the analysis requires isolation of plasma cells (especially in samples with <20% plasma cells), which is time-consuming, and requires significant expertise in interpretation of signal patterns. Primary cytogenetic abnormalities include hyperdiploidy and translocations involving the immunoglobulin heavy chain (*IGH*) locus on chromosome 14q32 (e.g., t(4;14), t(11;14) and t(14;16)). Translocations involving the *IGH* gene are primary (initiating) aberrations in PCM and are present in up to 60% of cases. Acquired secondary cytogenetic abnormalities may arise and include deletions of 17p as well as gains of 1q and deletions of 1p. Certain cytogenetic abnormalities are associated with a positive response to specific therapies (e.g., venetoclax with t(11;14); *IGH-CCND1*; bortezomib with t(4;14); *IGH-FGFR3/MMSET*). Neoplastic plasma cells are also present in the circulation and can be detected by sensitive flow cytometric methods (Nowakowski et al., 2005; Sanoja-Flores et al., 2018).

Immuno-flowFISH has been successfully applied to detect numerical and structural chromosomal abnormalities in multiple myeloma. Plasma cells were identified by CD38 and CD138 co-expression, a pattern that is highly specific for plasma cells; T cells and bone marrow precursor cells were CD38 positive and negative for CD138 expression. A range of FISH probes have been analyzed and abnormalities detected have included trisomies (chromosomes 11 and 14), monosomy 17, del(17p), t(4;14) and t(11;14). Immuno-flowFISH has been capable of identifying FISH abnormalities in both bone marrow and circulating plasma cells. One report describes monosomy 17, with one spot for both centromere 17 and 17p12 probes, being detectable in CD38, CD138 co-expressing cells in both sample types (Fig. 4) with as few as 0.01% positive blood cells. This exquisite sensitivity was achieved due to the large cell number analyzed and the specificity achieved through antigenic identification of the plasma cells (Erber et al., 2020; Lam et al., 2022). Chromosomal translocations involving the *IGH* gene (14q32) have been successfully detected by immuno-flowFISH using dual-fusion FISH probes for t(4;14); *IGH-FGFR3* and t(11;14); *IGH-MYEOV*. The FISH patterns seen with translocations are comparable to conventional FISH-on-a-slide, i.e., adjacent red and green FISH signals indicating a fusion (Fig. 4). To confirm these are true fusions, and not coincident, requires the presence of one intact green (1G) and red (1R) FISH signal for the uninvolved chromosomes. In addition, a cut-off (2%) should be used to exclude chance adjacent signal, a figure calculated from having normal cells in the analysis (i.e., T lymphocytes). A "1G1R2F" pattern is therefore indicative of balanced reciprocal chromosomal translocations.

3.3 Low-level disease assessment

The presence of residual disease and circulating neoplastic cells are helpful prognostic and predictive tools that can inform clinical decision-making. To date, this has not been possible using standard FISH techniques due to small numbers of cells assessed and the low test sensitivity (i.e., one positive cell in 20). The precision of immuno-flowFISH offers a new testing paradigm for minimal residual disease detection and detecting rare circulating cells, something not achievable with standard FISH. The phenotypic "gating" identifies the cells even when present in small numbers. This enables one genotypically abnormal cell to be detected in 10^5 cells, the current limit and which is at least four-logs greater than manual slide-based FISH methods. Evidence of this sensitivity has come from studying CLL with low level disease detected with both trisomy 12 and del(17p) (Fuller et al., 2021). This sensitivity has also been reported with myeloma where circulating plasma cells with monosomy 17 could be seen at a level of 0.01% of 300,000 blood leukocytes analyzed (Lam et al., 2022). This was achieved because of the CD38/CD138-gating highlighting the cell of interest. This was performed on whole blood (after red cell lysis); no prior plasma cell isolation step was used. An additional application of the technology for detecting small numbers of cells includes analysis for emergent subclones which may indicate disease relapse or progression and be at low level.

The success of this approach could lead to a strategy for real-time blood-based whole-cell FISH analysis for monitoring and to detect clonal evolution. Since this is new technology, thresholds will need to be developed to define a clinically-relevant positive FISH result that correlates with adverse clinical outcomes, such as relapse or disease progression. Further, the emergence of a new FISH-detectable chromosomal lesion will be a useful finding, regardless of burden, as this could indicate clonal evolution and potential progression to more aggressive disease.

This flow cytometric testing paradigm for blood cancers has potential to influence the practice for other cancers with aneuploidy (e.g., melanoma; breast cancer) and which shed tumor cells into the bloodstream. Such cells are actively being pursued as targets for cancer detection and characterization (i.e., "liquid biopsy"). Most liquid biopsy approaches are hindered by the rarity of the cancer cells in the blood, an inability to differentiate the cancer cells from blood cells even if present in sufficient numbers, and limitations on the diagnostic information available from the cancer cells following their identification. These limitations can potentially be overcome using immuno-flowFISH and suitable antibodies for the tumor type (e.g., epithelial antigens). Circulating tumor cells (CTCs) have been demonstrated to be useful biomarkers for MRD monitoring of a number of cancers, with the number correlating with tumor stage, outcome and therapeutic response (Letestu et al., 2021; Sanoja-Flores et al., 2018; van Dongen, van der Velden, Brüggemann, & Orfao, 2015). Surveillance for CTCs with chromosomal changes could be applied to identify relapsed disease or emergent new clones with new genomic defects and thereby enable early clinical intervention. This will potentially bring orders-of-magnitude improvements in diagnostic sensitivity to a broad range of cancers and detection of circulating neoplastic cells based on their abnormal chromosomal makeup.

4. Conclusions

Chromosomal abnormalities can be detected in leukemias by imaging flow cytometry using new generation immuno-flowFISH technology. This automated high precision FISH technique improves the sensitivity and specificity of current methodologies through positive identification of the neoplastic cell of interest and assessing it for chromosomal abnormalities. This new frontier for assessing cytogenetic aberrations in whole cells in suspension by imaging flow cytometry is an evolving field which opens new opportunities in diagnostic hematology and potentially other fields of oncology. Prior to real-time adoption, other hematological malignancies, antibodies and chromosomal defects need to be studied and the limits of detection standardized. The technology and applications will evolve alongside advances in the instrumentation and analytical capabilities from machine learning for data interpretation.

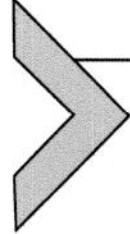

5. Methods/protocol/technical appendix

5.1 Materials

5.1.1 Blood and bone marrow samples

Blood collected in 9 mL Vacuette K3 Ethylenediaminetetraacetic acid (EDTA) Tube (Greiner, cat# 455036)

Bone marrow collected in 4 mL Vacuette K2 Ethylenediaminetetraacetic acid (EDTA) Tube (Greiner, cat# 454023)

5.1.2 Reagents

BD PharmLyse solution (dilute 10× stock 1:10 in MilliQ water) (BD Biosciences, cat# 555899)

15 mL centrifuge tube (Greiner, cat# 188271)

1× phosphate buffered saline (PBS) at room temperature (RT) and ice-cold: dilute 5 mL 10×PBS in 45 mL MilliQ water

Hemacytometer (Merck, cat# Z359629)

0.4% Trypan blue solution (Thermo Fisher Scientific, cat# 15250061)

5 mL Eppendorf tubes (Eppendorf, cat# 0030119460)

2% Fetal bovine serum (FBS; Bovogen Biologicals, cat# SFBSNZ) in 1×PBS
MilliQ water, MilliQ Advantage A10 water purification system (Merck, cat# Z00Q0V0WW)
BS3 cross-linking solution (see Buffers and solutions)
Quench buffer: 5 mL of 1 M Tris-hydrochloride and 45 mL of 150 mM sodium chloride (see Buffers and solutions)
4% Formaldehyde with 0.1% Tween20 buffer (see Buffers and solutions)
0.5 M HCl (see Buffers and solutions)
Vysis CEP hybridization buffer (Abbott Molecular, cat# 07J36-001)
Vysis LSI/WCP hybridization buffer (Abbott Molecular, cat# 06J67-001)
0.1% Igepal-CA630 in 2×SSC (see Buffers and solutions)
1.5 mL lo-bind microtubes (Sigma-Aldrich, cat# T4816)
0.2 mL PCR tubes (Qiagen, cat# 981005)
0.3% Igepal-CA630 in 0.4×SSC, prewarmed to 42 °C (see Buffers and solutions)
SYTOX AADvanced working solution (see Buffers and solutions)
Sphero Rainbow Calibration Particles (8 Peaks), 3.0–3.4 μm (BD Biosciences, cat# 559123)

5.1.3 Antibody list

Antibody (clone)	Isotype	Fluorophore	Catalog #	Manufacturer	Malignancy
CD3 (SK7)	IgG1, κ	BV605	563217	BD Biosciences	CLL
CD5 (UCHT2)	IgG1, κ	AF647	300616	Biolegend	CLL
CD19 (SJ25C1)	IgG1, κ	BV480	566103	BD Biosciences	CLL
CD38 (HB-7)	IgG1, κ	AF647	356632	Biolegend	Myeloma
CD138 (MI15)	IgG1, κ	BV480	566140	BD Biosciences	Myeloma
Mouse isotype (X40)	IgG1, κ	BV480	565652	BD Biosciences	CLL, Myeloma
Mouse isotype (X40)	IgG1, κ	BV605	562652	BD Biosciences	CLL
Mouse isotype (MOPC-21)	IgG1, κ	AF647	400130	Biolegend	CLL, Myeloma

5.1.4 FISH probe list

Probe	Fluorophore	Catalog #	Manufacturer	Malignancy
Vysis CEP12	SpectrumGreen	06J36-022	Abbott Molecular	CLL
Vysis CEP17 (D17Z1)	SpectrumGreen	06J37-017	Abbott Molecular	CLL
SureFISH 17p12	OrangeRed	G101179R-8	Agilent Technologies	CLL
CytoCell IGH/FGFR3 Translocation, Dual Fusion	IGH-FITC FGFR3-TexasRed	LPH 030	Oxford Gene Technology	Myeloma
CytoCell IGH/ MYEOV Translocation, Dual Fusion	IGH-FITC MYEOV-TexasRed	LPH 045	Oxford Gene Technology	Myeloma

5.1.5 Instruments

- Vortex
- Centrifuge with swing buckets
- Microscope
- Microfuge
- 96-well thermal cycler
- Heat block
- Amnis® ImageStream®X Mark II with Amnis® IDEAS® v6.3 Image Analysis Software (Luminex Corporation, Seattle, USA)

5.2 Buffers and solutions

BS3 stock solution (20 mM)

- 4 mg bis(sulfosuccinimidyl)suberate (BS3) (Sigma-Aldrich, cat# S5799)
- 350 μL 1.25×PBS
- Store up to 1 year at −20 °C

BS3 cross-linking solution (1 mM)

- 40 μL 20 mM BS3 stock solution
- 760 μL 1.25×PBS
- Use within 1 h, discard unused solution

4% Formaldehyde with 0.1% Tween20

- 250 mL Pierce 16% formaldehyde (*w/v*) methanol-free solution (Thermo Fisher Scientific, cat# 28908)
- 740 μL 1×PBS
- 10 μL 10% Tween20
- Buffer should be made at RT and combined for 10 min prior to use
- Discard unused buffer

0.5 M hydrochloric acid (HCl)

- 209 μL 12 M/37% HCl SG1.18 (VWR Chemical, cat# 20252.420)
- 1 mL MilliQ water
- Adjust to pH 0.7
- Add MilliQ water to make 5 mL
- Prepare solution in fume cabinet
- Add HCl to water slowly

10×PBS (stock solution)

- 160 g sodium chloride (Sigma-Aldrich, cat# S9888)
- 4 g potassium chloride (Ajax Chemical, cat# AJA383)
- 28.8 g di-Sodium hydrogen orthophosphate (Na_2HPO_4) (Ajax Chemical, cat# AJA621)
- 4.8 g potassium dihydrogen orthophosphate (KH_2PO_4) (Fischer Chemical, cat# P/4800/50)
- 1.6 L MilliQ water
- Dissolve solutes in MilliQ water over low heat using a magnetic stirrer
- Adjust pH to 7.4, make up to final volume of 2 L with MilliQ water
- Filter, autoclave and store up to 1 year at RT

1.25×PBS

- 6.25 mL 10×PBS
- 43.75 mL MilliQ water
- Store up to 6 months at RT

150 mM sodium chloride

- 0.4383 g sodium chloride (Sigma-Aldrich, cat# S9888)
- 50 mL MilliQ water
- Store up to 1 year at RT

20×SSC (3 M sodium chloride/0.3 M sodium citrate) stock solution

- 8.766 g sodium chloride (Sigma-Aldrich, cat# S9888)
- 4.9 g sodium citrate (Chem-Supply, cat# SA034)
- 50 mL MilliQ water
- Adjust to pH 7.0.
- Store up to 1 year at RT

2×SSC

- 1 mL 20×SSC
- 9 mL MilliQ water
- Store up to 1 year at RT

0.4×SSC

- 200 μL 20×SSC
- 9 mL MilliQ water
- Store up to 1 year at RT

SYTOX AADvanced stock solution (1 μM)

- 1 μL SYTOX AADvanced (Thermo Fisher Scientific, cat# S10349)
- 999 μL 1×PBS
- Store up to 1 year at −20 °C.

SYTOX AADvanced working solution (0.2 μM)

- 20 μL SYTOX AADvanced stock solution
- 80 μL 1×PBS
- Use within 1 h, discard unused solution

1 M Tris-hydrochloride (Tris-HCl)

- 6.075 g Tris-HCl (Sigma-Aldrich, cat# T6066)
- 35 mL MilliQ water
- pH to 7.4, make up to 50 mL in MilliQ water
- Store up to 1 year at RT

10% Tween20

- 100 μL Tween20 (Sigma-Aldrich, cat# 9005-64-5)
- 900 μL 1×PBS
- Solution is viscous, pipette slowly and mix well
- Store up to 1 year at RT

10% Igepal-CA630

- 500 μL Igepal-CA630 (Sigma-Aldrich, cat# 9002-93-1)
- 4.5 mL 2×SSC
- Solution is viscous, mix well
- Store up to 2 weeks at RT

0.1% Igepal-CA630 in 2×SSC

- 10 μL 10% Igepal-CA630
- 9.99 mL 2×SSC
- Store up to 2 weeks at RT

0.3% Igepal CA630 in 0.4×SSC

- 30 μL 10% Igepal-CA630
- 9.97 mL 0.4×SSC
- Store up to 2 weeks at RT

5.3 Protocol

Immunophenotyping, fixation and permeabilization

1. Add 1 mL of EDTA-anticoagulated blood or bone marrow to 10 mL of 1 × BD PharmLyse and gently vortex. *Note*: Process sample within 24 h of collection
2. Incubate at room temperature (RT), protected from light, for 10 min
3. Centrifuge 200 × *g* for 5 min, remove supernatant
4. Add 5 mL PBS, centrifuge 200 × *g* for 5 min, remove supernatant. Repeat wash
5. Resuspend in 250 μL–2 mL PBS for cell counting. *Note*: Adjust volume to 5 mL for patients if leukocyte count is $>2 \times 10^{11}$/L
6. Remove 10 μL of the cell suspension, dilute with 90 μL of trypan blue and count cells using a hemocytometer chamber. Calculate volume to add $2–5 \times 10^6$ cells per test based on calculated cell concentration (cells/mL)
7. Aliquot $5–10 \times 10^6$ viable cells to each sample in a 5 mL Eppendorf tube. *Note*: Assessment should include test sample (stained/probed), unstained, isotype control, and single stained or single probed as compensation controls
8. Add 600 μL 2%FBS/PBS, centrifuge at 900 × *g* for 3 min, remove supernatant
9. Add monoclonal antibodies (5 μL/1×10^6 cells) or isotypes to cell pellet, resuspend thoroughly, incubate 30 min at 4 °C. *Note*: Samples should be protected from light for all remaining protocol steps
   a. CLL panel: CD3-BV605, CD5-AF647 and CD19-BV480
   b. Multiple myeloma panel: CD38-AF647 and CD138-BV480
10. Add 800 μL 2%FBS/PBS, centrifuge 900 × *g* for 3 min, remove supernatant
11. Resuspend completely in 200 μL BS3 cross-linking solution, incubate 30 min at 4 °C
12. Add 1 mL Quench buffer, incubate 20 min at 4 °C. *Note*: Add buffer slowly to the side of the tube. Do not add buffer directly to cells and do not aspirate cells
13. Add 800 μL 2%FBS/PBS, do not aspirate or resuspend cells when adding 2%FBS/PBS. Centrifuge 900 × *g* for 3 min, remove supernatant
14. Fix samples by adding 250 μL of 4% Formaldehyde with 0.1% Tween20 buffer to a loosened cell pellet, resuspend thoroughly with aspiration, incubate 10 min at RT
15. Add 800 μL 2%FBS/PBS, centrifuge 900 × *g* for 3 min, remove supernatant

16. Resuspend in 800 μL 2%FBS/PBS. *Note*: An aliquot of each test sample can be removed at this step for analysis with isotype controls to confirm immunophenotyping
17. Centrifuge 900×*g* for 3 min, remove supernatant

DNA denaturation and fluorescence in situ hybridization

18. Resuspend fully in 100 μL 0.5 M HCl, incubate 20 min at RT
19. Quench in 3 mL ice-cold PBS, centrifuge 600×*g* for 10 min, remove supernatant
20. Prepare FISH probe mix in a 0.2 mL lo-bind PCR tube and pre-warm at 37 °C for 10 min
    a. CLL: add 7 μL Vysis CEP hybridization buffer, 1 μL CEP12-SG probe and 2 μL MilliQ water; or add 7 μL Vysis LSI/WCP hybridization buffer, 1 μL CEP17-SG probe, 1 μL SureFISH 17p12-OR probe and 1 μL MilliQ water
    b. Multiple myeloma: add 10 μL IGH/FGFR3 or 10 μL IGH/MYEOV probe mix
21. After quench block samples in 1 mL 2%FBS/PBS, centrifuge 900×*g* 3 min, remove supernatant.
22. Add 150 μL 0.1% Igepal/2xSSC stringency wash buffer, transfer cells to 0.2 mL lo-bind PCR tube, centrifuge cells at 950×*g* for 3 min
23. Remove supernatant and all excess buffer from samples in PCR tubes. *Note*: It is important all wash buffer is removed as this may change the ratio of the hybridization buffer and affect hybridization efficiency
24. Resuspend cells in the pre-warmed probe mix
25. Place samples in thermal cycler with heated lid, set reaction volume setting to 10–30 μL to ensure even heating, denature at 78 °C for 5 min then hybridize at 42 °C for at least 24 h.

Stringency wash and nuclear staining

26. Add 150 μL 0.1% Igepal/2×SSC stringency wash buffer, aspirate gently to mix sample, transfer to 1.5 mL lo-bind microfuge tube
27. Centrifuge 950×*g* for 3 min, remove supernatant.
28. Resuspend in 200 μL 0.3% Igepal/0.4×SSC stringency wash buffer (pre-warmed to 42 °C), incubate for 5 min at 55 °C to degrade excess probe
29. Add 800 μL 2%FBS/PBS, centrifuge 950×*g* for 3 min, remove supernatant
30. Resuspend in 30 μL SYTOX AADvanced working solution to stain DNA, incubate 20 min at RT

Imaging flow cytometry

31. Set the following acquisition parameter values on a 2-camera 12-channel AMNIS ISX MkII imaging flow cytometer:

 a. Illumination: 405 nm laser at 100 mW, 488 nm laser at 200 mW, 561 nm laser at 200 mW, 642 nm laser at 120 mW, SSC laser (785 nm) at 1.5 mW, Brightfield ON for Channel 1 and 9;
 b. Magnification and EDF: 60× objective with extended depth of field (EDF) ON;
 c. Fluidics: set to "lo" speed and "hi" sensitivity.

32. Load Sphero Rainbow Calibration particles prepared with manufacturer's instructions. Acquire 1000 events on gate set on bead population in Area vs Aspect Ratio plot. Display bead population in fluorescence intensity histograms to establish 6–8 peak histogram profiles to ensure reproducible instrument performance
33. Load test samples and acquire from 10,000 to 500,000 events from gate set for single cells in a Brightfield Area vs Aspect Ratio plot. On average, each sample can be acquired at a rate of 1–300 cells per second, i.e., 1–10 min/10,000 event collection for each sample to ensure uniformity and reproducibility in the collection of single cell events.
34. Acquire data for 1000–5000 cells in unstained and isotype controls
35. Open the Compensation Wizard and acquire data for 1000 cells in single stained compensation controls

IDEAS® imaging flow cytometry data analysis

36. Analyze data in IDEAS® v6.1 image analysis software. *Note*: The optimized IDEAS analysis templates associated to each disease application described below can be saved as a template for assessment of subsequent samples to standardize workflow and data reproducibility and to generate reliable interpretation of the chromosomal data of neoplastic cells (i.e., generate reference ranges and appropriate in-house laboratory cut-off values)
37. Select Guided Analysis > Wizards then select the test sample .rif file for analysis
38. Create a compensation matrix with single stained compensation controls
39. Set image display properties to include:
 a. CLL: Brightfield (Ch01), SpectrumGreen FISH probe (Ch02), OrangeRed FISH probe (Ch03), SYTOX AADvanced (Ch05), side scatter (Ch06), BV480 (Ch07), BV605 (Ch10), and AF647 (Ch11)
 b. Multiple myeloma: Brightfield (Ch01), FITC FISH probe (Ch02), TexasRed FISH probe (Ch04), SYTOX AADvanced (Ch05), side scatter (Ch06), BV480 (Ch07), and AF647 (Ch11)
40. Select Begin Analysis Wizard, use the Gradient Root Mean Square (RMS) feature histogram to select Brightfield (Ch01) images in focus

(Fig. 2A). *Note*: RMS measures the sharpness quality of an image by detecting large changes in pixel values in the image, cells with better focus have higher Gradient RMS values

41. Gate single cells with a scatter plot of the Brightfield Area versus Aspect Ratio (Fig. 2B). Exclude cell doublets, debris and clumps. *Note*: The Aspect Ratio features the Minor Axis of the object or cell divided by the Major Axis and describes how circular or oblong an object is where single cells have a high Aspect Ratio and low Area value

42. Gate nucleated nondividing cells using a SYTOX AADvanced fluorescence intensity histogram (Fig. 2C). Exclude cells with high fluorescence intensity (dividing cells, dead cells and cell clumps

43. Gate single nucleated non-dividing cells with hybridized FISH probe signals (Fig. 2D and E):

   a. CLL: gate single nucleated cells with hybridized CEP12 probe signals with a bivariate plot of CEP12-SG probe max pixel versus CEP12-SG probe fluorescence intensity; or gate single nucleated cells with hybridized CEP17 probe signals with a bivariate plot of CEP17-SG probe max pixel vs CEP17-SG probe fluorescence intensity, then gate single nucleated CEP17 hybridized cells with hybridized LSI 17p12 probe signals with a bivariate plot of 17p12-OR probe max pixel versus 17p12-OR probe fluorescence intensity and exit wizard

   b. Multiple myeloma: gate single nucleated cells with hybridized IGH probe signals with a bivariate plot of IGH-FITC probe max pixel versus IGH-FITC probe fluorescence intensity, then gate single nucleated IGH hybridized cells with hybridized FGFR3 or MYEOV probe signals with a bivariate plot of FGFR3-TxR or MYEOV-TxR probe max pixel versus FGFR3-TxR or MYEOV-TxR probe fluorescence intensity and exit wizard

44. Gate cell populations based on immunophenotype (Fig. 2F):

   a. CLL: generate a scatterplot of fluorescence intensity of CD19 (B-lymphocytes including CLL) vs CD3 (normal T-lymphocytes) and CD5 (CLL and normal T-lymphocytes) and gate the $CD19^{+}$ $CD3^{-}CD5^{+}$ CLL population.

   b. Multiple myeloma: generate a scatterplot of fluorescence intensity of CD38 vs CD138 and gate the $CD38^{+}CD138^{+}$ myeloma plasma cell population.

45. To count FISH probe spots using the Spot Count feature calculation wizard select Guided Analysis > Wizards > Spot wizard. For each probe select the focus and single cell populations defined above. Click Yes to "Do you want to analyze sub-populations" question and select the cell population defined above based on immunophenotype. Create truth populations for low (1–2 spots) and high (4+ spots). *Note*: The automated Spot Count Feature algorithm that examines the connectivity of each pixel based on whether it is connected to a particular spot or the background and enumerates the number of FISH "spots" in each cell for an entire sample. The "Spot" mask for each probe can be calibrated and the Spot count feature function can be appended with the "Intensity" mask for greater precision in spot-counting performance for samples with greater background in the mask manager. Select Function > Spot > Mask: Bright (M02 and Ch02 for SpectrumGreen or FITC, M03 and Ch03 for SpectrumOrange or OrangeRed and M04 and Ch04 for TexasRed)

a. Defining the "Spot" mask function: set minimum radius of "0" and a maximum radius of "4," then adjust "Spot to Cell Background Ratio" from a range of 1.00–30.00 (increasing ratio). This ratio is the spot pixel value divided by the mean background value in the bright detail image. For this adjustment, pay attention to the software defined blue masked area as you adjust. The mask will get smaller as the ratio is increased, the final value should produce a mask that precisely or tightly overlays with the observed fluorescent spots in the probe channel

b. Defining the "Intensity" mask function: append the "And" operator after the "Spot" mask definition to select and incorporate the "intensity" function. Set "maximum" intensity to 4095 (or max value) and adjust the "minimum" intensity, paying attention to the software defined blue masked area, which will get smaller as the minimum intensity is increased. This can be assisted by the software by "pointing the cursor" over individual pixels of the probe/cell image to determine specific FISH pixel intensity values to calibrate the intensity limit. The final value should produce a mask that precisely or tightly overlays the observed fluorescent spots in the probe channel (from 170 to 600). Press "OK" to complete function and confirm overall definition for updated software masks to begin recalculating spot counts

46. IDEAS will calculate the Spot Count feature values and graph. Click finish to close the wizard and gate the 0, 1, 2, 3, and 3+ spot count increments on the x-axis to enumerate percentage of cells in each category (Fig. 2G)

47. Repeat Spot Count wizard calculations for each probe in assay (Fig. 2H)
48. Repeat steps 44–47 for each population in sample (Fig. 2I–K):
    a. CLL: plot CD3 (T-lymphocytes) vs CD5 (CLL and T-lymphocytes) for the population in step 43 and gate the $CD3^{+}CD5^{+}$ T-lymphocyte population
    b. Multiple myeloma: on the scatterplot of CD38 vs CD138, gate the $CD38^{-}CD138^{-}$ population
49. Fluorescence adjustment of CEP probes can be performed with single parameter histograms comparing the measured fluorescence intensity of FISH signals for each of the spot count populations (i.e., 1-spot vs 2-spot vs 3-spot). Changes in measured fluorescence intensity correspond to the number of hybridized probes, allowing correction of overlapping signals (i.e., 2-spots that appear as 1-spot) inherent in 2-dimensional image projections to determine the true number of specific FISH spots and reliable distinction of monosomy and disomy subpopulations. This analysis can also be used to verify the presence of true trisomy events (e.g., +12) by measuring the higher fluorescence intensity of three specific FISH signals hybridized to the centromere as opposed to an extra non-specific probe signal. Fluorescence adjustment is usually performed on centromeric probes (i.e., CEP12 or 17) as the current platform is limited in the ability to resolve and quantify (based on intensity) small locus specific probe signals (<600kb).
50. Software generated spot counts can be verified in the image gallery to further refine the gate strategy. An "overlay" or merged image of the immunophenotype, SYTOX DNA stain and probe FISH signals can also be viewed to confirm specificity of all probe signals (i.e., localization to the nucleus of each cell) and composite FISH signal patterns (e.g., 2G2R).

Analysis of CLL for trisomy 12 and del(17p):

1. Assess the number of FISH spots per cell for each population. Normal B and T lymphocytes will have 2 chromosome 12 enumeration probe signals (2 CEP12 spots, disomy 12); CD19+/CD5+ CLL cells with trisomy 12 will have 3 spots for CEP12; and CD19+/CD5+ CLL cells with del(17p) will have 2 CEP17 signals and a single co-localized chromosome 17p12 signal (2 green spots and 1 yellow spot/ 2G1Y)
2. Calculate "spot count ratios" from the spot count statistics for each subpopulation and each probe. This is achieved by calculating mean spot count of CLL cells divided by the mean spot count of normal T-lymphocytes:

$$\frac{\text{CLL}\ (\text{CD19}^{+}\text{CD5}^{+}\text{CD3}^{-})}{\text{T}-\text{lymphocytes}\ (\text{CD19}^{-}\text{CD5}^{+}\text{CD3}^{+})}$$

Note: Normal T-lymphocytes will have a mean spot count of close to two for CEP12, CEP17 and 17p region. A CLL/T-cell spot ratio of <1 or >1 will indicate loss or gain of chromosomal regions under investigation respectively. Loss of 17p will have a mean 17p12 spot count of less than 2 and CLL/T ratio of <1 (e.g., del(17p)). Cells with a ratio of >1 may indicate trisomy (e.g., trisomy 12).

Analysis of multiple myeloma for t(4;14); IGH-FGFR3 and t(11;14); IGH-MYEOV:

1. Use the image gallery to assess the FISH probe fluorescence and SYTOX AADvanced merged images for the $\text{CD38}^{+}\text{CD138}^{+}$ population with 3 FISH spots and confirm spatial proximity of the signals. Note: The presence of *IGH* structural abnormalities involving *FGFR3/MYEOV* partners resulted in 3-spot signal patterns for each *IGH* (G) and *FGFR3* or *MYEOV* (R) due to "splitting" of one locus involved in a balanced reciprocal chromosomal translocation
2. Identify fusion (F) of translocated signals by the presence of two green (G) and red (R) FISH spots in close proximity or "fused," and the presence of single independent FISH spots for both probes or defined as "1G1R2F." A 1G1R2F" FISH pattern represents t(4;14); *IGH-FGFR3* and t(11;14); *IGH-MYEOV* in the plasma cells of multiple myeloma
3. Use the image gallery to assess the FISH probe fluorescence and SYTOX AADvanced merged images for the $\text{CD38}^{-}\text{CD138}^{-}$ population and in healthy controls to identify normal "2G2R" diploid FISH signals and establish cut-offs to discriminate chance chromosome overlap from real fused signals due to chromosomal translocations.

Conflict of interest statement

H.Y.L.H., W.N.E. and K.A.F. have filed Australian (AU2018355889) and international (US20200232019, CN111448324, P2018870567) patents related to the immunoflowFISH protocol.

References

Barteneva, N. S., Fasler-Kan, E., & Vorobjev, I. A. (2012). Imaging flow cytometry: Coping with heterogeneity in biological systems. *The Journal of Histochemistry and Cytochemistry*, *60*(10), 723–733. https://doi.org/10.1369/0022155412453052.

Basiji, D. A., Ortyn, W. E., Liang, L., Venkatachalam, V., & Morrissey, P. (2007). Cellular image analysis and imaging by flow cytometry. *Clinics in Laboratory Medicine*, *27*(3), 653–670. viii https://doi.org/10.1016/j.cll.2007.05.008.

Cambier, J. L., Kay, D. B., & Wheeless, L. L., Jr. (1979). A multidimensional slit-scan flow system. *The Journal of Histochemistry and Cytochemistry*, *27*(1), 321–324. https://doi.org/10.1177/27.1.374595.

Erber, W. N., Hui, H., Stanley, J., Mincherton, T., Clarke, K., Augustson, B., et al. (2020). Detection of del(17p) in hematological malignancies by imaging flow cytometry. *Blood*, *136*(Supplement 1), 9–10. https://doi.org/10.1182/blood-2020-143323.

Fuller, K. A., Bennett, S., Hui, H., Chakera, A., & Erber, W. N. (2016). Development of a robust immuno-S-FISH protocol using imaging flow cytometry. *Cytometry. Part A*, *89*(8), 720–730. https://doi.org/10.1002/cyto.a.22852.

Fuller, K., Hui, H., Stanley, J., & Erber, W. N. (2021). FISH by imaging flow cytometry in CLL for diagnosis and MRD assessment. *Blood*, *138*, 2619. https://doi.org/10.1182/blood-2021-152266.

Grimwade, L. F., Fuller, K. A., & Erber, W. N. (2017). Applications of imaging flow cytometry in the diagnostic assessment of acute leukaemia. *Methods*, *112*, 39–45. https://doi.org/10.1016/j.ymeth.2016.06.023.

Grimwade, L., Gudgin, E., Bloxham, D., Bottley, G., Vassiliou, G., Huntly, B., et al. (2012). Detection of cytoplasmic nucleophosmin expression by imaging flow cytometry. *Cytometry. Part A*, *81*(10), 896–900. https://doi.org/10.1002/cyto.a.22116.

Grimwade, L., Gudgin, E., Bloxham, D., Scott, M. A., & Erber, W. N. (2011). PML protein analysis using imaging flow cytometry. *Journal of Clinical Pathology*, *64*(5), 447–450. https://doi.org/10.1136/jcp.2010.085662.

Hui, H. Y. L., Clarke, K. M., Fuller, K. A., Stanley, J., Chuah, H. H., Ng, T. F., et al. (2019). "Immuno-flowFISH" for the assessment of cytogenetic abnormalities in chronic lymphocytic leukemia. *Cytometry. Part A*, *95*(5), 521–533. https://doi.org/10.1002/cyto.a.23769.

Hui, H., Fuller, K. A., Chuah, H., Liang, J., Sidiqi, H., Radeski, D., et al. (2018). Imaging flow cytometry to assess chromosomal abnormalities in chronic lymphocytic leukaemia. *Methods*, *134-135*, 32–40. https://doi.org/10.1016/j.ymeth.2017.11.003.

Hui, H. Y. L., Stanley, J., Clarke, K., Erber, W. N., & Fuller, K. A. (2021). Multi-probe FISH analysis of immunophenotyped chronic lymphocytic leukemia by imaging flow cytometry. *Current Protocols*, *1*(10), e260. https://doi.org/10.1002/cpz1.260.

Kachel, V., Benker, G., Lichtnau, K., Valet, G., & Glossner, E. (1979). Fast imaging in flow: A means of combining flow-cytometry and image analysis. *The Journal of Histochemistry and Cytochemistry*, *27*(1), 335–341. https://doi.org/10.1177/27.1.374598.

Lam, S., Mincherton, T. I., Hui, H. Y. L., Sidiqi, M. H., Fuller, K. A., & Erber, W. N. (2022). Imaging flow cytometry shows monosomy 17 in circulating plasma cells in myeloma. *Pathology*, *54*(7), 951–953.

Letestu, R., Dahmani, A., Boubaya, M., Baseggio, L., Campos, L., Chatelain, B., et al. (2021). Prognostic value of high-sensitivity measurable residual disease assessment after front-line chemoimmunotherapy in chronic lymphocytic leukemia. *Leukemia*, *35*(6), 1597–1609. https://doi.org/10.1038/s41375-020-01009-z.

Minderman, H., Humphrey, K., Arcadi, J. K., Wierzbicki, A., Maguire, O., Wang, E. S., et al. (2012). Image cytometry-based detection of aneuploidy by fluorescence in situ hybridization in suspension. *Cytometry Part A.*, *81A*(9), 776–784. https://doi.org/10.1002/cyto.a.22101.

Nowakowski, G. S., Witzig, T. E., Dingli, D., Tracz, M. J., Gertz, M. A., Lacy, M. Q., et al. (2005). Circulating plasma cells detected by flow cytometry as a predictor of survival in 302 patients with newly diagnosed multiple myeloma. *Blood*, *106*(7), 2276–2279. https://doi.org/10.1182/blood-2005-05-1858.

Sanoja-Flores, L., Flores-Montero, J., Garcés, J. J., Paiva, B., Puig, N., García-Mateo, A., et al. (2018). Next generation flow for minimally-invasive blood characterization of MGUS and multiple myeloma at diagnosis based on circulating tumor plasma cells (CTPC). *Blood Cancer Journal*, *8*(12), 117. https://doi.org/10.1038/s41408-018-0153-9.

Scarfo, L., Ferreri, A. J., & Ghia, P. (2016). Chronic lymphocytic leukaemia. *Critical Reviews in Oncology/Hematology*, *104*, 169–182. https://doi.org/10.1016/j.critrevonc.2016.06.003.

Thompson, P. A., Tam, C. S., O'Brien, S. M., Wierda, W. G., Stingo, F., Plunkett, W. S., et al. (2016). Fludarabine, cyclophosphamide, and rituximab treatment achieves long-term disease-free survival in IGHV-mutated chronic lymphocytic leukemia. *Blood*, *127*(3), 303–309. https://doi.org/10.1182/blood-2015-09-667675.

van Dongen, J. J. M., van der Velden, V. H. J., Brüggemann, M., & Orfao, A. (2015). Minimal residual disease diagnostics in acute lymphoblastic leukemia: Need for sensitive, fast, and standardized technologies. *Blood*, *125*(26), 3996–4009. https://doi.org/10.1182/blood-2015-03-580027.

CHAPTER FIVE

Efficient discrimination of functional hematopoietic stem cell progenitors for transplantation by combining alkaline phosphatase activity and CD34$^+$ immunophenotyping

Laura G. Rico[a], Jordi Juncà[a], Roser Salvia[a], Michael D. Ward[b], Jolene A. Bradford[b], and Jordi Petriz[a,*]

[a]Functional Cytomics Lab, Germans Trias i Pujol Research Institute (IGTP), ICO-Hospital Germans Trias i Pujol, Universitat Autònoma de Barcelona, Badalona, Barcelona, Spain
[b]Thermo Fisher Scientific, Fort Collins, CO, United States
*Corresponding author: e-mail address: jpetriz@igtp.cat

Contents

Methods in Cell Biology, Volume 195
ISSN 0091-679X
https://doi.org/10.1016/bs.mcb.2023.08.003

Abstract

Alkaline phosphatase (ALP) is a membrane-associated hydrolase enzyme with dimeric structure that catalyzes phosphate esters, optimally at alkaline pH. ALP has a focus of interest, since this enzyme is highly expressed in primitive stem cells, such as progenitor cells, non-differentiating cells, and primordial cells. We previously adapted a fluorescent microscopy-based assay for quantifying ALP^{high} and ALP^{low} cells by flow cytometry in combination with immunophenotyping. Our method uses a minimal sample perturbation approach, avoiding the use of erythrocyte lysing solutions and washing steps, and offering opportunities to combine live cell response and functional assessment with cell immunophenotyping, while minimizing sample preparation effects on the cell biology. Here we provide a detailed experiment protocol to determine alkaline phosphatase activity in $CD34^{+}$ hematopoietic stem cells from blood and apheresis products obtained from patients involved in a stem cell mobilization process for allo- or auto-transplant. This study may provide the early detection of progenitors at different levels of differentiation and therefore, relate this information to long-term engraftment in hematopoietic stem cell transplants.

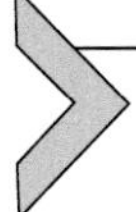

1. Introduction

1.1 Alkaline phosphatase

Alkaline phosphatase (ALP) is a membrane-associated hydrolase enzyme with dimeric structure that catalyzes phosphate esters, optimally at alkaline pH. In humans, there are three isoenzymes restricted to a specific tissue: (1) intestinal ALP (Hass, Wada, Herman, & Sussman, 1979), (2) placental ALP (Terao & Mintz, 1987), and (3) germ cell ALP (Berstine, Hooper, Grandchamp, & Ephrussi, 1973); and a nonspecific tissue ALP isoenzyme, found in bone, liver, and kidney (McKenna, Hamilton, & Sussman, 1979; Seargeant & Stinson, 1979). In the study of stem cell biology, ALP has a focus of interest, since this enzyme is highly expressed in primitive stem cells, such as progenitor cells, non-differentiating cells and primordial cells (Berstine et al., 1973; Takahashi, Tanabe, Ohnuki, et al., 2007). Moreover, ALP is highly expressed during development, indicating cell differentiation potential and stemness.

Alkaline phosphatase activity was originally detected by using a technique described by Takamatsu and Akahoshi (1956) and Gomori and Benditt (1953). This histochemical technique was based on the deposition of calcium phosphate at the site of enzyme action. Later, techniques such as ELISA, western blot, and immunohistochemistry in combination with chromogenic substrates were commonly used to assay ALP levels on different cells. More recently, highly sensitive fluorescent and chemiluminescent substrates have been developed to analyze alkaline phosphatase activity in both fixed and living cells (Singh et al., 2012).

In previous studies of our group (Bardina et al., 2020; Rico, Juncà, Ward, Bradford, & Petriz, 2016; Rico, Juncà, Ward, Bradford, & Petriz, 2019), we adapted a fluorescent microscopy-based assay for quantifying ALP^{high} and ALP^{low} cells by flow cytometry in combination with immunophenotyping. Our method uses a minimal sample perturbation approach, avoiding the use of erythrocyte lysing solutions and washing steps. This strategy offers opportunities to combine live cell response and functional assessment with cell immunophenotyping, while minimizing sample preparation effects on the cell biology.

1.2 Hematopoietic stem cell transplant

Hematopoietic stem cell transplantation (HSCT) involves the intravenous infusion of autologous or allogeneic $CD34^+$ cells to reestablish hematopoietic function (Copelan, 2006). HSCT can involve peripheral blood leukapheresis to collect $CD34^+$ cells from autologous or allogeneic donors (Padmanabhan, 2018; Panch, Szymanski, Savani, & Stroncek, 2017). This procedure may be the only curative option in many hematological malignant and nonmalignant conditions, and it is generally safe and well-tolerated (Copelan, 2006).

1.3 Rationale

The aim of the present study was to use this rapid quantitative assay for detection of ALP activity in $CD34^+$ hematopoietic cells before and after stem cell collection, as a potential biomarker for short- and long-term engraftment.

Here we provide a detailed experiment protocol to determine alkaline phosphatase activity in $CD34^+$ hematopoietic stem cells obtained from blood and apheresis products from patients involved in a stem cell mobilization process for allo- or auto-transplantation.

2. Materials

2.1 Biological samples

Mobilized peripheral blood and apheresis products used in this study are obtained from healthy donors and multiple myeloma patients. Blood samples are collected in EDTA-anticoagulated tubes after hematopoietic cell progenitor mobilization using granulocyte colony stimulating factors (G-CSF) are analyzed the same day of the extraction to preserve their functional and biological characteristics. Apheresis products are obtained after a leukapheresis procedure and are analyzed on the same day of the extraction. Samples are collected from patients visiting Hospital Universitari Germans Trias i Pujol (HUGTIP) (Barcelona, Spain).

All patients enrolled in this study provided their informed consent following the Declaration of Helsinki. All procedures are under the internal protocols of our laboratory, which are authorized by the HUGTIP Clinical Investigation Ethical Committee, in agreement with current Spanish legislation.

2.2 Disposables

1. EDTA anticoagulated blood collection tubes
2. 1.5 ml microcentrifuge tubes
3. 12x75 mm polypropylene tubes
4. Pipettes and pipette tips

2.3 Equipment

1. Tube rotator
2. Hemocytometer, or another cell counting device
3. Laboratory water bath set at exactly 37 °C
4. Flow Cytometer with the appropriate software, e.g., Attune NxT™ Flow Cytometer (Thermo Fisher, cat. no. A24858) with Attune NxT™ software v. 3.1.1162.1 (Thermo Fisher)
5. Filter kit to detect Violet Side Scatter, e.g., Attune NxT™ No-Wash No-Lyse Filter Kit (Thermo Fisher, cat. no. 100022776)

2.4 Reagents and solutions

1. Hanks' Balanced Salt Solution, calcium- and magnesium-free, without phenol red (HBSS; Capricorn Scientific GmbH, cat. no. HBSS-2A)
2. Fetal Bovine Serum (FBS; Biowest, cat. no. S18B-500)
3. Vybrant™ DyeCycle™ Violet Stain 5 mM in Dimethyl Sulfoxide (DMSO) (DCV; Invitrogen™, cat. no. A14353)
4. Alkaline Phosphatase Live Stain (APLS; Thermo Fisher, cat. no. A14353)
5. PE-CD34, clone AC136 (Miltenyi Biotec, cat. no. 130-113-179)

3. Sample preparation

1. Mobilized peripheral blood samples obtained in EDTA anticoagulated tubes and apheresis products are prepared immediately, ideally in less

than 4 h after extraction. To preserve samples, they are maintained in a tube rotator at room temperature.

2. For sample preparation, a circulating water bath is set at 37 °C and APLS, stored at –20 °C, is removed to allow it to thaw.
3. Then, nucleated cells are counted accurately using a hemocytometer or another cell counting device.
4. The required sample volume to obtain 0.5 to 1.0×10^6 cells is prepared in a microcentrifuge tube. HBSS is added to reach 100 µL final volume. In high cellular samples (e.g., apheresis products) it is recommended to pre-dilute the sample to avoid pipetting small volumes. In any case, reverse pipetting is highly recommended.

4. DNA staining, blockading, and Alkaline Phosphatase Live Staining

1. 1 µL DCV 5 mM is added to the microcentrifuge tube at a final concentration of 50 µM to stain nucleated cells.
2. 10 µL FBS are added to the microcentrifuge tube to block non-specific binding.
3. 1 µL APLS 0.1 mM (defrosted) is added to the microcentrifuge tube at a final concentration of 1 µM to determine alkaline phosphatase activity in live cells.
4. Sample is resuspended carefully and incubated for 20 min at 37 °C in a dedicated water bath protected from light.

5. Immunophenotyping

1. After incubation, 5 µL PE-CD34 at a final concentration of 5 µg/µL are added to identify hematopoietic stem cells.
2. Sample is resuspended carefully and incubated for 20 min at room temperature protected from light.

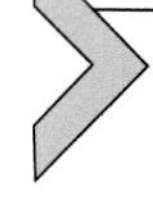

6. Flow cytometric acquisition and data analysis

6.1 Flow cytometer configuration for sample acquisition

In this chapter, we have collected data on the Attune NxT™ Flow Cytometer (Thermo Fisher). This flow cytometer is equipped with acoustic-assisted hydrodynamic-focusing and 4 lasers (405 nm violet,

488 nm blue, 561 nm yellow/green and 638 nm red). The lasers, detectors, filter configuration, and scale of each used parameter are listed on Table 1. Attune™ No-Wash No-Lyse Filter Kit (Thermo Fisher) allows a configuration capable of collecting Side Scatter (SSC) with the violet laser to increase scatter resolution when acquiring unlysed samples. Violet side scatter is more accurate than blue side scatter for whole blood analysis. Nevertheless, a default blue SSC signal can also be used in case the cytometer does not allow a violet SSC configuration. For the acquisition of whole blood samples, the height parameter is more accurate than the area parameter because the area parameter is affected by erythrocytes and background coincidence.

1. After monoclonal antibody incubation, samples are diluted with 2 mL HBSS and transferred to a 12 × 75 mm polypropylene tube. Once prepared, samples must be analyzed immediately by flow cytomety.
2. Samples are acquired at 25–100 μL/min, assuring an event rate no higher than 400 events/s. To have significant results, a minimum of 100,000 nucleated cells (defined by DCV-positive events) are acquired.
3. Threshold levels are set empirically in the DCV channel, by using a violet SSC versus DCV dual plot to eliminate debris and the considerable number of erythrocytes that are found in unlysed peripheral blood for detection. When analyzing leukapheresis products, threshold levels are also set in the DCV channel.

6.2 Gating strategy and analysis of flow cytometry data

Flow cytometry files are analyzed with flow cytometry software. Results of this chapter are presented with Attune™ NxT Software v.3.1.1162.1 (Thermo Fisher) and FlowJo Software v.10.8 (Becton Dickinson). Fig. 1 shows a representative workspace to determine alkaline phosphatase activity on CD34$^+$ hematopoietic stem cells.

1. First, a dual density plot displaying V-SSC-H versus DCV-H is created to discriminate non-nucleated cells (erythrocytes, platelets), debris, and some necrotic cells from nucleated cells (Fig. 1A).
2. The "Nucleated cells" region is gated on a dual density plot displaying DCV-H vs. DCV-A for doublets discrimination by fluorescence (Fig. 1B).
3. The "Single cells" region is gated on a dual FSC-H versus SSC-H density plot used to check scatter distribution (Fig. 1C).

Table 1 Attune™ NxT Flow Cytometer configuration.

Parameter	Laser	Detector	Pulse parameter	LP filter	DLP filter	BP filter	Scale
FSC	488 nm	FSC	Height	Blank	No filter	488/10 nm	Linear
Blue SSC	488 nm	SSC	Height	Blank	495 nm	488/10 nm	Linear
Violet SSC	405 nm	VL1	Height	Blank	415 nm	405/10 nm	Linear
DCV	405 nm	VL2	Height and Area	413 nm	495 nm	440/50 nm	Logarithmic
APLS	488 nm	BL1	Height	496 nm	555 nm	530/30 nm	Logarithmic
PE-CD34	561 nm	YL1	Height	569 nm	600 nm	585/16 nm	Logarithmic

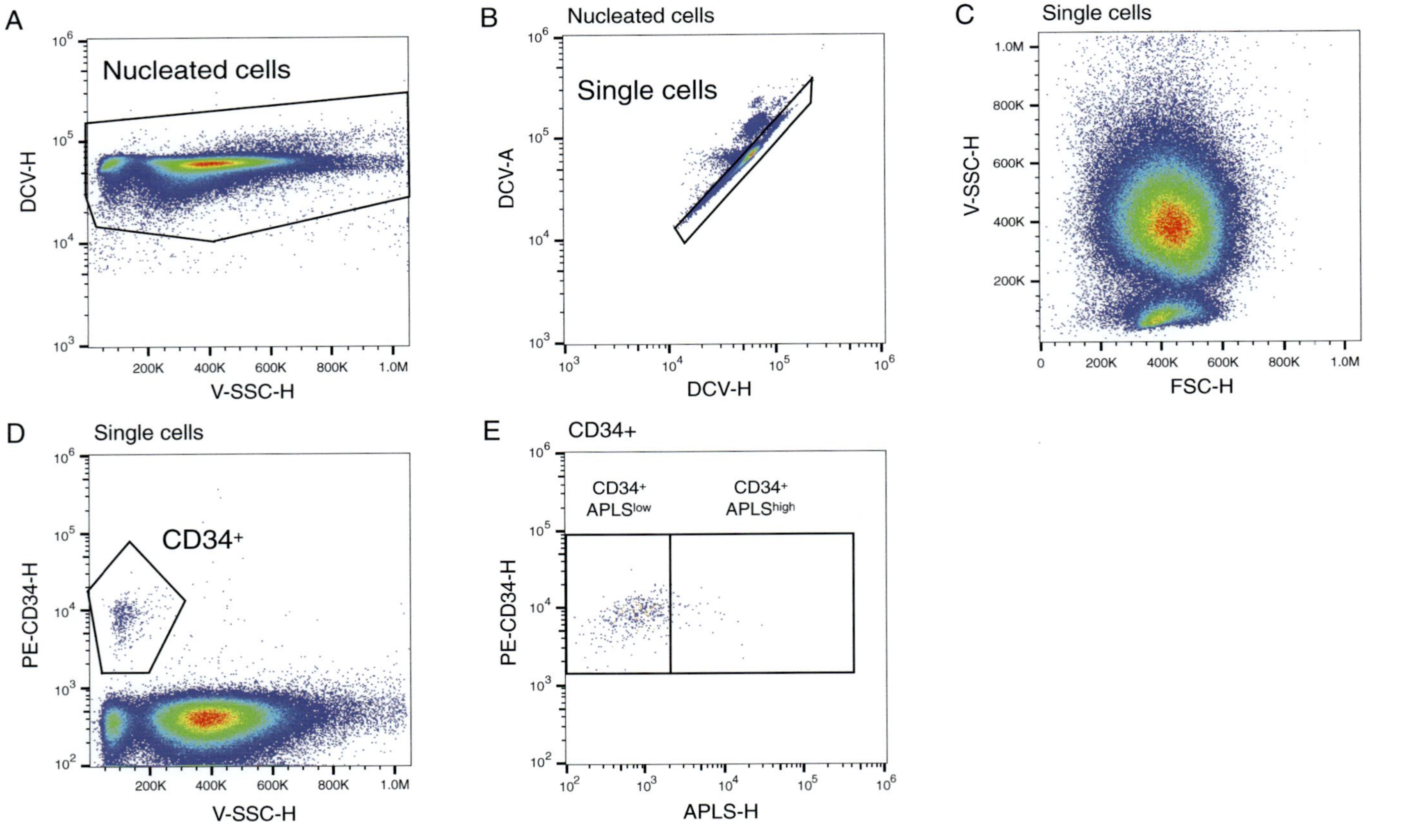

Fig. 1 Representative flow cytometry workspace to determine alkaline phosphatase activity on CD34$^+$ hematopoietic stem cells. A dual density plot displaying V-SSC-H versus DCV-H is created to discriminate non-nucleated cells (erythrocytes, platelets), debris, and some necrotic cells from nucleated cells (A). The nucleated cells region is gated on a dual density plot displaying DCV-H versus DCV-A for doublets discrimination by fluorescence (B). The single cells region is gated on a dual FSC-H versus SSC-H density plot used to check scatter distribution (C). The single cells region is gated on a dual V-SSC-H versus PE-CD34-H density plot to select the CD34$^+$ population (D). CD34$^+$ region is gated on an APLS-H versus PE-CD34-H density plot to determine CD34$^+$APLS^{high} cells and CD34$^+$APLS^{low} cells (E). Data was collected on the Attune™ NxT Flow Cytometer (Thermo Fisher). FCS files were analyzed with FlowJo Software v.10.8 (Becton Dickinson).

4. The "Single cells" is gated on a dual V-SSC-H versus PE-CD34-H density plot (Fig. 1D) to show the fluorescence of the monoclonal antibody used and select the CD34$^+$ population.
5. Finally, the "CD34$^+$" region is gated on an APLS-H versus PE-CD34-H density plot to determine CD34$^+$APLS^{high} cells and CD34$^+$ APLS^{low} cells (Fig. 1E).

6.3 Statistical analysis

Prism 9 (GraphPad Prism Sofware, Dotmatics) is used to perform statistical analysis. Kruskall-Wallis test is used to compare variance differences between multiple variables and Mann-Whitney test is used to compare variance differences between two variables. A P-value <0.05 is considered significant.

7. Representative results

We initially designed a flow cytometry protocol to detect and quantify alkaline phosphatase activity in CD34$^+$ cells from mobilized peripheral blood and apheresis products obtained in $n = 20$ healthy donors (HD) and $n = 20$ multiple myeloma (MM) patients. Fig. 1 shows the representative workspace designed to quantify alkaline phosphatase activity on CD34$^+$ hematopoietic stem cells. We observed different levels of alkaline phosphatase activity in CD34$^+$, as represented in Fig. 2. HD showed significantly higher CD34$^+$ cells concentration than MM patients in both mobilized peripheral blood (Fig. 3A, MM day-1 vs. HD P-value <0.0001; MM day 0 vs. HD P-value$=0.0089$) and apheresis products (Fig. 3B, MM vs. HD P-value$=0.0016$). Regarding CD34$^+$ with high alkaline phosphatase activity, we also found higher concentration in HD in both mobilized peripheral blood (Fig. 3C, MM day-1 vs. HD P-value <0.0001; MM day 0 vs. HD P-value$=0.0199$) and apheresis products (Fig. 3D, MM vs. HD P-value$=0.0022$).

7.1 Concluding remarks

CD34$^+$ cells have been shown to have differential levels of alkaline phosphatase activity at the cellular level. It has been described that the higher the activity of this enzyme, the greater the undifferentiation, as demonstrated by fluorescence microscopy on embryonic stem cells (González,

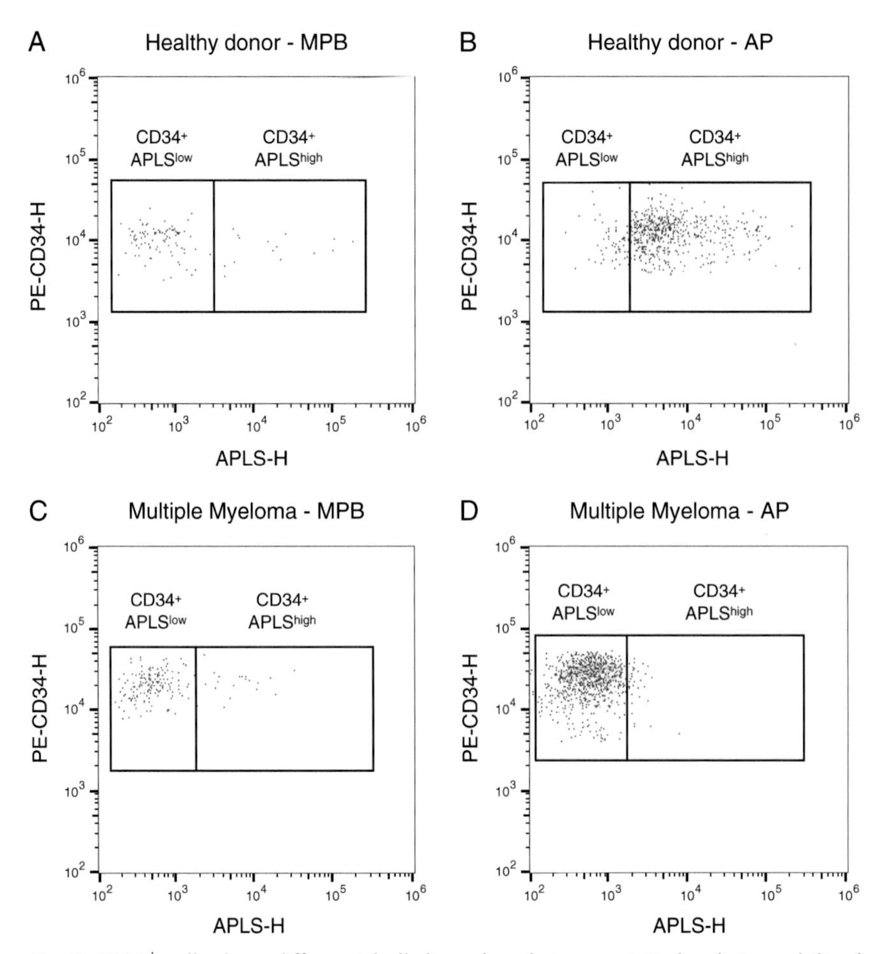

Fig. 2 CD34$^+$ cells show differential alkaline phosphatase activity levels in mobilized peripheral blood (MPB) and apheresis products (AP). CD34$^+$ alkaline phosphatase levels of activity detected in a mobilized peripheral blood from a healthy donor (HD) obtained before apheresis (A) and in an apheresis product (B). CD34$^+$ alkaline phosphatase levels of activity detected in a mobilized peripheral blood from a multiple myeloma (MM) patient obtained before apheresis (C) and in an apheresis product (D). Data was collected on the Attune™ NxT Flow Cytometer (Thermo Fisher). FCS files were analyzed with FlowJo Software v.10.8 (Becton Dickinson).

Boué, & Belmonte, 2011). In the case of mobilized peripheral blood, it is possible to detect higher levels of circulating hematopoietic progenitors after mobilization (Padmanabhan, 2018), allowing the application of the methodology and quantification of alkaline phosphatase activity. The leukapheresis products show even higher numbers of these progenitor cells

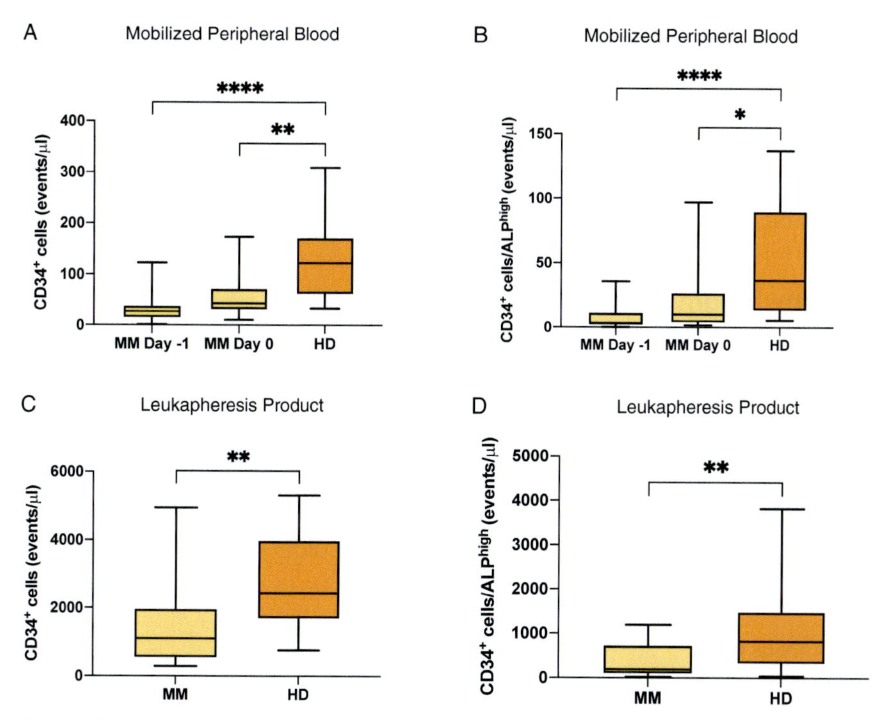

Fig. 3 Enumeration of CD34$^+$ cells and CD34$^+$ cells with high levels of Alkaline Phosphatase activity (ALP^{high}) in healthy donors (HD) and multiple myeloma (MM) patients. (A) CD34$^+$ cells absolute number (events/μL) detected in mobilized peripheral blood from MM patients ($n = 20$) before apheresis (day −1 and day 0) and HD ($n = 20$) before apheresis (day 0). (B) Absolute number (events/μL) of CD34$^+$ cells with high alkaline phosphatase activity detected in mobilized peripheral blood from multiple myeloma patients before apheresis (day −1 and day 0) and healthy donors before apheresis (day 0). (C) Absolute number of CD34$^+$ cells (events/μL) detected in apheresis products from MM patients and HD. (D) Absolute number (events/μL) of CD34$^+$ cells with high alkaline phosphatase activity detected in apheresis products from MM patients and HD. Data was collected on the Attune™ NxT Flow Cytometer (Thermo Fisher). FCS files were analyzed with FlowJo Software v.10.8 (Becton Dickinson). The data obtained were analyzed and presented using Prism v9 (Graphpad Prism Software, Dotmatics).

just by concentrating mononuclear cells (Padmanabhan, 2018). It is precisely by using the product enriched in CD34$^+$ cells that it is possible to detect progenitors with differential levels of alkaline phosphatase activity, a phenomenon that could be related to different functional capacities of these cells. Hence, as hypothesized, we could be detecting progenitors at different levels of differentiation and therefore, relate this information to long-term engraftment. Comparative analyses between CD34$^+$ progenitors in peripheral blood mobilized from healthy donors and patients undergoing a transplantation procedure reveal the different quality of the hematopoietic product, and studies at the functional level based on flow cytometry and cell separation will be necessary to better understand the biology of these intriguing cells.

Acknowledgments

This study was partially supported by Thermo Fisher Scientific. We thank the CERCA Programme/Generalitat de Catalunya and Germans Trias i Pujol Research Foundation for institutional support and acknowledge financial support from the Obra Social la Caixa. The authors are grateful to Clara Streiff, Paola Paglia, Pierre Le Ninan, Marjorie Joiner, Stephane Behar, Sergio Ramon, Lluís Sainz, and Víctor Querol from Thermo Fisher Scientific for all their help in this research field, and to Sara Vergara and Minerva Raya for kindly providing samples used in this study. Part of this work was done while the corresponding author was principal investigator at the Josep Carreras Leukemia Research Institute (Spain).

Disclosures

M.D.W. and J.A.B. are employees of Thermo Fisher Scientific, which is in the business of selling flow cytometers and flow cytometry reagents. The rest of authors declare no potential conflicts of interest.

References

Bardina, J., Rico, L. G., Ward, M. D., Bradford, J. A., Juncà, J., & Petriz, J. (2020). Flow cytometric quantification of granulocytic alkaline phosphatase activity in unlysed whole blood. *Current Protocols in Cytometry*, *93*(1). https://doi.org/10.1002/cpcy.76.

Berstine, E. G., Hooper, M. L., Grandchamp, S., & Ephrussi, B. (1973). Alkaline phosphatase activity in mouse teratoma. *Proceedings of the National Academy of Sciences of the United States of America*, *70*(12), 3899–3903. https://doi.org/10.1073/pnas.70.12.3899.

Copelan, E. A. (2006). Hematopoietic stem-cell transplantation. *The New England Journal of Medicine*, *354*(17), 1813–1826. https://doi.org/10.1056/NEJMra052638.

Gomori, G., & Benditt, E. P. (1953). Precipitation of calcium phosphate in the histochemical method for phosphatase. *The Journal of Histochemistry and Cytochemistry*, *1*(2), 114–122. https://doi.org/10.1177/1.2.114.

González, F., Boué, S., & Belmonte, J. C. I. (2011). Methods for making induced pluripotent stem cells: Reprogramming à la carte. *Nature Reviews. Genetics*, *12*(4), 231–242. https://doi.org/10.1038/nrg2937.

Hass, P. E., Wada, H. G., Herman, M. M., & Sussman, H. H. (1979). Alkaline phosphatase of mouse teratoma stem cells: Immunochemical and structural evidence for its identity as a somatic gene product. *Proceedings of the National Academy of Sciences of the United States of America*, *76*(3), 1164–1168. https://doi.org/10.1073/pnas.76.3.1164.

McKenna, M. J., Hamilton, T. A., & Sussman, H. H. (1979). Comparison of human alkaline phosphatase isoenzymes. Structural evidence for three protein classes. *The Biochemical Journal*, *181*(1), 67–73. https://doi.org/10.1042/bj1810067.

Padmanabhan, A. (2018). Cellular collection by apheresis. *Transfusion, 58*, 598–604. https://doi.org/10.1111/trf.14502.

Panch, S. R., Szymanski, J., Savani, B. N., & Stroncek, D. F. (2017). Sources of hematopoietic stem and progenitor cells and methods to optimize yields for clinical cell therapy. *Biology of Blood and Marrow Transplantation, 23*(8), 1241–1249. https://doi.org/10.1016/j.bbmt.2017.05.003.

Rico, L. G., Juncà, J., Ward, M. D., Bradford, J., & Petriz, J. (2016). Is alkaline phosphatase the smoking gun for highly refractory primitive leukemic cells? *Oncotarget*, 7(44), 72057–72066. https://doi.org/10.18632/oncotarget.12497.

Rico, L. G., Juncà, J., Ward, M. D., Bradford, J. A., & Petriz, J. (2019). Flow cytometric significance of cellular alkaline phosphatase activity in acute myeloid leukemia. *Oncotarget, 10*(65), 6969–6980. https://doi.org/10.18632/oncotarget.27356.

Seargeant, L. E., & Stinson, R. A. (1979). Evidence that three structural genes code for human alkaline phosphatases. *Nature, 281*(5727), 152–154. https://doi.org/10.1038/281152a0.

Singh, U., Quintanilla, R. H., Grecian, S., Gee, K. R., Rao, M. S., & Lakshmipathy, U. (2012). Novel live alkaline phosphatase substrate for identification of pluripotent stem cells. *Stem Cell Reviews and Reports, 8*(3), 1021–1029. https://doi.org/10.1007/s12015-012-9359-6.

Takahashi, K., Tanabe, K., Ohnuki, M., et al. (2007). Induction of pluripotent stem cells from adult human fibroblasts by defined factors. *Cell, 131*(5), 861–872. https://doi.org/10.1016/j.cell.2007.11.019.

Takamatsu, H., & Akahoshi, Y. (1956). A new technique for the histochemical demonstration of phosphatase in hard tissue. *Acta Tuberculosea Japonica, 6*(1), 11–19. https://doi.org/10.1016/j.cell.2007.11.019.

Terao, M., & Mintz, B. (1987). Cloning and characterization of a cDNA coding for mouse placental alkaline phosphatase. *Proceedings of the National Academy of Sciences of the United States of America, 84*(20), 7051–7055. https://doi.org/10.1073/pnas.84.20.7051.

CHAPTER SIX

PD-L1 expression in multiple myeloma myeloid derived suppressor cells

Laura G. Rico[a,b], Roser Salvia[a,b], Jolene A. Bradford[c], Michael D. Ward[c], and Jordi Petriz[a,b,*]

[a]Functional Cytomics Lab, Germans Trias i Pujol Research Institute (IGTP), ICO-Hospital Germans Trias i Pujol, Universitat Autònoma de Barcelona, Badalona, Barcelona, Spain
[b]Department of Cellular Biology, Physiology and Immunology, Autonomous University of Barcelona (UAB), Cerdanyola del Vallès, Spain
[c]Thermo Fisher Scientific, Fort Collins, CO, United States
*Corresponding author: e-mail address: jpetriz@igtp.cat

Contents

Abstract

The Programmed Cell Death Protein 1/Programmed Cell Death Protein Ligand 1 (PD-1/PD-L1) axis stands as one of the most widely acknowledged targets for cancer immunotherapy. This ligand is considered a therapeutic target for this disease as it might play an important role in tumor immune evasion and drug resistance. In multiple myeloma

Methods in Cell Biology, Volume 195
ISSN 0091-679X
https://doi.org/10.1016/bs.mcb.2024.11.006

(MM), PD-L1 is overexpressed in abnormal plasma cells and Myeloid-Derived Suppressor Cells (MDSCs). In MDSCs, unlike tumoral cells or derived cell lines, the PD-L1 protein is presented in a conformation not recognized by the monoclonal antibody. In contrast, when stimulating the sample with PMA, the PD-L1 molecule undergoes a conformational change that enables its recognition. Hence, we have developed a flow cytometric screening assay to determine PD-L1 conformational changes in MDSCs based on a minimal manipulation of the sample, to preserve the structure and functionality of the ligand. In this chapter, we provide detailed protocols to assess PD-L1 levels in MDSCs together with the representative results obtained in multiple myeloma patients. The obtained results enable the classification of MM patients based on the different PD-L1 detection after stimulation, which increases compared with unstimulated samples. We also provide protocols to assess kinetic analysis of PD-L1 expression over time and to compare PD-L1 cell surface expression with cytoplasmic expression. Finally, competitive experiments in the presence of durvalumab are also described to study its interaction with PD-L1. This approach can also be used to study the contribution of potential conformational changes in other proteins.

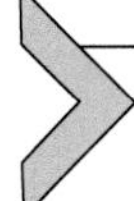

1. Introduction

1.1 Multiple Myeloma

1.1.1 Definition, characteristics, etiology and epidemiology

Multiple myeloma (MM) is the second most common hematological cancer, comprising 15–20% of all hematopoietic cancers and causing around 20% of all deaths from these diseases (Siegel, Miller, Fuchs, & Jemal, 2021). In MM, terminally differentiated plasma cells (PCs) are expanded and accumulated within the bone marrow (BM), constituting a BM microenvironment known to be critically important for MM survival, growth, and chemosensitivity. Malignant plasma cells typically secrete the monoclonal immunoglobulin protein (M protein), an aberrant pool of immunoglobulins (Kumar et al., 2017). The accumulation of M protein, the interaction of clonal plasma cells, and cytokine secretion result in hypercalcemia, renal insufficiency, anemia, and/or bone lesions, known as CRAB symptoms (Brigle & Rogers, 2017; Kumar et al., 2017). Although new therapeutic approaches developed during the last decades have improved patient outcomes, most MM patients show drug resistance and eventually relapse. Therefore, MM currently remains an incurable disease.

1.1.2 The immune checkpoint PD-L1/PD-1

It is known that the progression of multiple myeloma benefits from immunological escape. The immune checkpoint of Programmed Death Receptor-1 (PD-1) and Programmed Death-Ligand 1 (PD-L1) is reported to critically influence the advance of MM by suppressing the immune surveillance (Tamura et al., 2005).

PD-1 is a type I transmembrane protein induced in activated T cells (Ishida, Agata, Shibahara, & Honjo, 1992), B cells and myeloid cells (Freeman et al., 2000). It plays a role in downregulating immune responses to maintain a peripheral tolerance and avoid autoimmunity. In a cancer environment, PD-1 on lymphocytes binds to PD-L1 and PD-L2 (also known as B7-H1 and B7-DC, respectively) found in tumoral cells, impairing the TCR-signaling pathway and cytokine secretion, and eventually inhibiting the cytotoxic activity of T cells and promoting its apoptosis (Freeman et al., 2000). In a MM study in humans, bone marrow $CD8^+$ T cells presented an enhanced expression of PD-1 compared to peripheral blood $CD8^+$ T cells, and also higher than $CD8^+$ T cells from the MM precursors MGUS (Monoclonal Gammopathy of Undetermined Significance) or SMM (Smoldering MM) (Kwon et al., 2020), indicating that the pathogenesis of MM induces the upregulation of PD-1 in BM $CD8^+$ T cells for T cell exhaustion or apoptosis.

Conversely, PD-L1 ligand is basically present in resting lymphocytes and antigen presenting cells (Yamazaki et al., 2002). In the tumor microenvironment, cancer cells also express this ligand to downregulate the immune response. Expression of PD-L2, the second ligand of PD-1, is limited to DCs and macrophages after activation (Ishida et al., 2002; Latchman et al., 2001; Tseng et al., 2001). In one hand, there is contradictory data about the levels of PD-L1 expression in both normal and myeloma plasma cells. One theory stands that PD-L1 is not expressed on plasma cells from neither MGUS nor healthy plasma cells, while the myeloma PCs have increased expression of PD-L1 when compared to the individuals (Dhodapkar et al., 2015; Kwon et al., 2020). On the other hand, other studies have shown that normal PCs express decreased levels of the ligand and that there is no significant difference in PD-L1 expression between aberrant or healthy PCs (Paiva et al., 2015). In addition, the levels of PD-L1 transcripts were found to be highly heterogeneous in MM patients but equivalent to those in normal PCs, indicating that the protein expression may be post-transcriptionally regulated (Kelly et al., 2018).

1.1.3 Molecular dynamics of PD-L1

PD-L1 identification has been a challenging topic for years and considerably discrepant data has been published to date, still not being resolved completely. Molecular dynamic studies imply that the PD-L1 molecule has a rather rigid conformation. It has been postulated that the major structural displacements are associated with the movement of the C″D loop, even though the structural conformation does not change significantly upon PD-1 binding because some loops adopt similar conformations in both free and antibody-bound PD-L1 structure (Ahmed & Barakat, 2017; Lee, Lee, & Heo, 2019). However, in another study, the authors propose a model of conformational transition for the binding of PD-L1 to PD-1 based on a delayed dissociation of the ligand from the receptor which, in addition, followed the same behavior with its alternative ligand, CD80. A longer PD-L1/PD-1 binding implies a stronger interaction because of more high-affinity conformations, and consequently a higher dissociation time (Ghiotto et al., 2010).

Besides conformational changes, PD-L1 has been described to be strongly glycosylated, presenting 45 kDa glycosylated and 33 kDa deglycosylated enzymatically. Glycosylation increases PD-L1 stabilization, and it allows a better PD-1 recognition and immunosuppressive function (Li et al., 2016, 2018). Despite PD-L1 glycosylation favoring the PD-1 binding, at the same time it impairs the binding of monoclonal antibodies to PD-L1, leading to an inaccurate recognition of the PD-L1 molecule (Lee, Lee, et al., 2019).

1.1.4 The myeloid-derived suppressor cells

The PD-L1 ligand can be further expressed in other kind of cells, such as the Myeloid-Derived Suppressor Cells (MDSCs), or in exosomes, and this additional presence can also play a role in the response of the disease.

MDSCs are a heterogeneous population of immature myeloid cells that are abnormally expanded in pathological conditions such as cancer, infection, inflammation, or traumatic stress (Gabrilovich & Nagaraj, 2009). Phenotypically, human MDSCs lack cell-specific surface markers and can differ upon tumor type. They are generally described as $CD11b^{+}$ $CD33^{+}$ $HLADR^{-/lo}$ cells, which can be further classified in $CD14^{+}/CD15^{-}$ monocytic MDSCs (M-MDSC) or $CD14^{-}/CD15^{+}$ polymorphonuclear MDSCs (PMN-MDSCs) (Filipazzi et al., 2007; Görgün et al., 2013; Zea et al., 2005). These two MDSCs subpopulations have distinct phenotype, morphology, function, and immunosuppressive capacity. Binsfeld et al. described that the neutrophil-like PMN-MDSC subset acquires a pro-angiogenic role in MM due to expression changes induced by myeloid cells. Additionally, PMN-MDSCs have a higher immunosuppressive potential than M-MDSCs in MM. Contrarily, M-MDSCs subpopulation presents a monocytic phenotype and morphology, and contains osteoclast precursors (Binsfeld et al., 2016; Görgün et al., 2013).

In MM, MDSCs are remarkably increased in peripheral blood and in the bone marrow, and they accumulate in the BM, where myeloma cells promote their survival and proliferation (Görgün et al., 2013). In turn, MDSCs directly induce myeloid cell proliferation and immunosuppression of both innate and adaptive immune response, by its secretion of suppressive cytokines (TGF-β, IL-10, IL-6), and PD-L1 and PD-L2 expression (Dysthe & Parihar, 2020; Ramachandran et al., 2013). MDSCs levels correlate with the prognosis and advance of the disease (Görgün et al., 2013; Van Valckenborgh et al., 2012; Wang et al., 2015) and are key for the effectiveness of PD-L1 immunotherapies based on its PD-L1 expression and prominent role in MM.

Exosomes should also be considered in the global picture of PD-L1/PD-1 immunotherapy, since tumor-derived exosomes contain PD-L1, presented both on the surface and within the particles, which can recapitulate the effect of cell-surface PD-L1 (Yin et al., 2021). In breast cancer, it has been described that exosomal PD-L1 can bind to PD-1 and induce a tumoral escape response from T cells (Yang et al., 2018). In a mouse model of MM, Wang et al. described that MDSCs uptake exosomes derived from BM-stromal cells, inducing MDSCs expansion in vitro and increasing the immunosuppression on T cells through the secretion of nitric oxide (Wang et al., 2015).

1.1.5 Immunotherapy to target the PD-1/PD-L1 checkpoint

Immunotherapy is a major approach used in the treatment of MM, even though it is limited because of its lack of long-lasting functional anti-myeloma T cell responses (Minnie & Hill, 2020). Specifically, the immunotherapy against MM consists of immunomodulatory imide drugs (IMiDs), analogs of thalidomide); monoclonal antibodies (anti-CD38); CAR-T cells; oncolytic virotherapies; and immune checkpoint inhibitors (α-PD-1/α-PD-L1); among other strategies (Soekojo, Ooi, de Mel, & Chng, 2020). So far, no drug targeting PD-1 nor PD-L1 has been already approved for MM but there are many clinical trials being conducted, using pembrolizumab and nivolumab for targeting PD-1, and durvalumab and atezolizumab for targeting PD-L1. Notably, in many studies, these monoclonal antibodies are administered in combination with IMiD therapy (lenalidomide or pomalidomide). Unfortunately, many of these clinical trials have been terminated because the combination of these agents has derived in toxicity and no significant improvement was obtained in the objective response rates (Badros et al., 2017; Mateos et al., 2019; Zafar Usmani et al., 2019) or prematurely suspended as a preventive measure

(Bar et al., 2020). On the contrary, other studies have been completed and reported a complete response, such as the pembrolizumab administration early after autologous hematopoietic cell transplantation followed by lenalidomide (D'Souza et al., 2019). According to the U.S. National Library of Medicine, currently there are some active clinical trials evaluating pembrolizumab and durvalumab in multiple myeloma patients. The distinct success of these checkpoint inhibitors between patients and cancers is not entirely understood; hence, further research in multiple myeloma treatment focusing in the α-PD-1/PD-L1 immunotherapy is needed to determine the potential of these drugs, alone or in combination with other agents.

1.2 Rationale

The determination of the PD-L1 levels in cancer patients is crucial to better describe their clinical point, to predict the outcome of anti-PD-L1 treatment, and to identify a putative measurable residual disease (MRD). For the assessment of the PD-L1 ligand, the analysis of the MDSC population in peripheral blood or bone marrow may constitute a representative method to evaluate the PD-L1 levels of the patients and to predict therapy response. We previously described a lung cancer case study where the stimulation with phorbol 12-myristate 13-acetate (PMA) induced a much higher reactivity to PD-L1 of MDSC than without stimulation ((Rico, Aguilar Hernández, et al., 2021)). In the present study, we provide methodology to determine PD-L1 levels in MDSCs, specifically in a cohort of MM patients selected at different clinical points. This methodology is based on a minimal manipulation of the sample and by mimicking the tumor microenvironment through PMA stimulation. This methodology enables the classification of myeloma patients based on the different PD-L1 detection after stimulation. Our results show that PMA increased the identification of the PD-L1 molecule from 1 to 650 times compared with unstimulated samples. Interestingly, PMA may mimic TCR signaling by activating protein kinase C and NF-κB pathways (Iwata, Ohoka, Kuwata, & Asada, 1996; Xue et al., 2017) among many other pleiotropic effects, although in MDSC its effects have yet to be elucidated. In addition, as PMA stimulation is performed during a brief period of time, cell immunophenotyping will not be affected. In this study, we also demonstrate that PD-L1 was not translocated from the cytoplasm to cell surface because it was not detected in the cytosol when compared with other markers such as CD11b. Therefore, PMA stimulation may induce a conformational change in PD-L1 with an important impact in anti-PD-L1 antibody reactivity.

In this chapter, we provide detailed protocols to assess PD-L1 levels in MDSCs together with the obtained representative results. First, we studied the PD-L1 MDSCs expression using a flow cytometric screening assay based on a minimal sample perturbation and stimulation with PMA. In order to design and validate the PD-L1 screening protocol by flow cytometry, $n=35$ marrow MM specimens were used. Additionally, we assessed PD-L1 cytoplasmic levels in 11 of these samples. Finally, we performed additional approaches to evaluate PD-L1 molecular dynamics. In one hand, we explored PD-L1 expression changes over time after PMA stimulation at 0, 1, 5, 10 and 60 min. On the other hand, we designed a competition assay to evaluate the anti-PD-L1 monoclonal antibody binding capacity in the presence of durvalumab, an anti-PD-L1 molecule used for immunotherapy. All these competition binding assays were performed using bone marrow samples obtained from multiple myeloma patients ($n=4$).

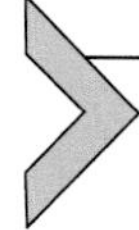

2. Materials and methods

2.1 Biological samples

Bone marrow samples used in this study were obtained from patients diagnosed with multiple myeloma, monoclonal gammopathy of undetermined significance (MGUS) and primary amyloidosis. Samples were obtained at different clinical points and analyzed the same day of the extraction to preserve their functional and biological characteristics. Samples were collected in EDTA-anticoagulating tubes from a series of patients attended at the Hospital Universitari Germans Trias i Pujol (HUGTIP) (Badalona, Spain).

All patients enrolled in this study provided their informed consent following the Declaration of Helsinki. All procedures were under the internal protocols of our laboratory, which were authorized by the HUGTIP Clinical Investigation Ethical Committee, in agreement with current Spanish legislation.

2.2 Reagents and solutions

- Hanks' Balanced Salt Solution, calcium- and magnesium-free, without phenol red (HBSS; Capricorn Scientific GmbH, cat. no. HBSS-2A).
- Fetal Bovine Serum (FBS; Biowest, cat. no. S18B-500)
- Dimethylsulfoxide (DMSO; Invitrogen™, cat. no. D12345)
- Vybrant™ DyeCycle™ Violet Stain (DCV; Invitrogen™, cat. no. A14353) 5 mM in DMSO.

- Phorbol 12-myristate 13-acetate (PMA, Sigma–Aldrich®, cat. no. P8139-1MG) 1 mg/mL in DMSO.
- Monoclonal antibodies:
  - PE-PD-L1, clone MIH1 (eBioscience™, cat. no. 12-5983-42)
  - PE-Cy7-CD33, clone WM53 (eBioscience™, cat. no. 25-0338-42)
  - FITC-HLA-DR, clone TU36 (Invitrogen™, cat. no. MHLDR01)
  - APC-CD11b, clone VIM12 (Invitrogen™, cat. no. CD11b05)
- Durvalumab; MEDI4736 (AstraZeneca, cat. no. ASZ4603)
- Ammonium chloride lysis solution 10x:
  - Distilled water
  - 1.5 M NH_4Cl; Ammonium chloride (Sigma-Aldrich®, cat. no. A9434)
  - 100 mM $NaHCO_3$; Sodium bicarbonate (Sigma-Aldrich®, cat. no. S5761)
  - 1 mM $C_{10}H_{14}N_2Na_2O_8$; EDTA, disodium salt (Calbiochem, cat. mo. 324,503)

    Prepared at pH = 7.4 (adjusted with 1 N HCl and 1 N NaOH)
- 70% Ethanol:
  - 72.6% EtOH, Neutral alcohol 93.6 ° (Alcoholes Gual S.A., cat. no. 64-17-5)
  - 27.4% Distilled water
- Hanks' Balanced Salt Solution with Albumin and Sodium Azide (HBA):
  - Hanks' balanced salt solution, calcium- and magnesium-free, without phenol red (HBSS; Capricorn Scientific GmbH, cat. no. HBSS-2A).
  - 1% Albumin, from bovine serum (Sigma-Aldrich®, cat. no. A7906)
  - 0.1% NaN_3; Sodium Azide extra pure (Sigma-Aldrich®, cat. no. 71290)

2.3 PD-L1 immunostaining

Regarding the study of PD-L1 expression, bone marrow specimens collected in EDTA-anticoagulated tubes are prepared accordingly minimal sample perturbation assays (Petriz, Bradford, & Ward, 2018; Rico, Salvia, Ward, Bradford, & Petriz, 2021). Importantly, samples should be prepared and acquired no more than 24 h after collection.

First, a marrow volume containing 10^6 nucleated cells is obtained and diluted to 100 μL with HBA buffer using 1.5 mL Eppendorf tubes and prepared per duplicate. Both tubes are incubated with 1 μL DCV (Final Concentration [FC] = 50 μM), and 10 μL FBS to block unspecific binding, at 37°C for 10 min in a dedicated water bath protected from light. After incubation, 900 μL HBA are added (1 mL final volume). One tube is used for cellular stimulation by adding 1 μL PMA (FC = 1.63 μM), and the second one is incubated with 1 μL of DMSO as a negative control. Samples are stimulated for 10 min at 37°C in a dedicated water bath protected from light. For kinetics study of PMA stimulation, different incubation times points are tested (e.g., 1, 5, 10 and 60 min).

After stimulation, tubes are centrifuged and 900 μL of supernatant are aspirated and reserved. Over the remaining 100 μL, cells are labeled by adding 5 μL of each of the following antibodies: PE-PD-L1, PE-Cy7-CD33, FITC-HLA-DR, and APC-CD11b, and incubated for 20 min at room temperature and protected from light. After incubation, labeled samples are resuspended using the reserved supernatant and analyzed immediately by flow cytometry.

2.4 Cytoplasmic PD-L1

In order to study PD-L1 at a cytoplasmic level, the necessary volume to obtain 1×10^6 nucleated cells is added in an Eppendorf tube. Erythrocytes are lysed using an ammonium chloride-based lysis solution for 10 min at room temperature. After that, samples are washed twice with HBSS and diluted in 1 mL 70% ethanol to fix and permeabilize cell membranes. Samples are incubated overnight at −20°C and washed twice with HBSS and resuspended in 100 μL HBA. Then, samples are incubated for 20 min at room temperature with 2 μL DAPI 1 mg/mL (FC = 6 μM), and 5 μL of PE-PD-L1 and APC-CD11b. CD11b is used as a positive control. After incubation, samples are diluted with HBA at a final volume of 1000 μL and acquired on the flow cytometry.

2.5 Competitive binding assays

To assess the potential crosstalk between the immunotherapy and the monoclonal antibody used for flow cytometry experimentation, a marrow volume containing 10^6 nucleated cells is obtained and diluted to 100 μL in HBA. Seven tubes are prepared and incubated with 1 μL DCV (FC = 50 μM) and 10 μL of FBS at 37°C for 10 min in a dedicated water bath protected from light. After incubation, 900 μL of HBA are added prior to cell stimulation. Then, 1 μL PMA (FC = 1.63 μM) is added to 6 tubes and the other one is incubated with 1 μL of DMSO (negative control). Cells are stimulated for 10 min at 37°C in a dedicated water bath protected from light.

After stimulation, tubes are centrifuged and 900 μL of supernatant are aspirated and reserved. Over the remaining 100 μL, cells are labeled by adding 5 μL of the following monoclonal antibodies: PE-PD-L1,

PE-Cy7-CD33, FITC-HLA-DR, and APC-CD11b. Durvalumab is then added to the different PMA stimulated tubes at increasing concentrations (0, 0.025, 0.25, 2.5, 25 and 250 μg/mL). All tubes are incubated for 20 min at room temperature and protected from light. After incubation, labeled samples are resuspended using the reserved supernatant and analyzed immediately by flow cytometry.

2.6 Flow cytometry

2.6.1 Flow cytometer configuration for sample acquisition

In this chapter, we have collected data on the Attune™ NxT Flow Cytometer (Thermo Fisher). This instrumentation is equipped with acoustic-assisted hydrodynamic-focusing and 4 lasers (405 nm violet, 488 nm blue, 562 nm yellow/green and 638 nm red). The lasers, detectors, filter configuration, and scale of each used parameter are listed on Table 1. Attune™ No-Wash No-Lyse Filter Kit (Thermo Fisher) is configured to collect Side Scatter with the violet laser to increase scatter resolution when acquiring unlysed samples (Petriz et al., 2018).

Samples are acquired at 25–100 μL/min using an event rate no higher than 400 events/s. To collect significant information, a minimum of 100,000 nucleated cells (defined by DCV) are acquired. Threshold levels are set empirically using a violet SSC (V-SSC) vs. DCV dual plot to eliminate debris and the large number of coincident erythrocytes.

Table 1 Attune™ NxT flow cytometer configuration.

Parameter	Laser	Detector	Pulse parameter	LP Filter	DLP Filter	BP Filter	Scale
FSC	488 nm	FSC	Height	Blank	555 nm	488/10 nm	Linear
Blue SSC	488 nm	SSC	Height	Blank	555 nm	488/10 nm	Linear
Violet SSC	405 nm	VL1	Height	Blank	495 nm	405/10 nm	Linear
DCV	405 nm	VL2	Height and Area	413 nm	495 nm	440/50 nm	Logarithmic
HLA-DR-FITC	488 nm	BL1	Height	496 nm	555 nm	530/30 nm	Logarithmic
7-AAD	488 nm	BL3	Height	496 nm	650 nm	695/70 nm	Logarithmic
PD-L1-PE	561 nm	YL1	Height	569 nm	600 nm	620/15 nm	Logarithmic
CD33-PE-Cy7	561 nm	YL4	Height	569 nm	740 nm	780/60 nm	Logarithmic
CD11b-APC	638 nm	RL1	Height	646 nm	690 nm	670/14 nm	Logarithmic

2.6.2 Gating strategy and analysis of flow cytometry data

Flow cytometry files are analyzed with the Attune™ NxT Software v.3.1.1162.1 (Thermo Fisher) and the FlowJo Software v.10.8 (Becton Dickinson). Fig. 1 shows a representative workspace to determine PD-L1 expression on MDSCs. First, a dual density plot displaying DCV-H versus V-SSC-H is created to discriminate non-nucleated cells (erythrocytes, platelets), debris, and some necrotic cells from nucleated cells (Fig. 1A). R1 is gated on a dual density plot displaying DCV-H vs. DCV-A for doublets discrimination by fluorescence (Fig. 1B). R2 is gated on a dual FSC-H versus SSC-H density plot used to finally discriminate necrotic cells and debris (R3) (Fig. 1C). R3 is gated on Fig. 1D–G, that represents FITC-HLA-DR (Fig. 1D), APC-CD11b (Fig. 1E), PE-PD-L1 (Fig. 1F), and PE-Cy7-CD33 (Fig. 1G) versus violet SSC to represent single fluorescence of the used monoclonal antibodies. Fig. 1H to L represent the combination of all fluorochromes to adjust color compensation of events gated on R3. Finally, Fig. 1M–O represent the MDSCs PD-L1+ gating strategy. These cells are selected as HLA-DR$^{low/neg}$/PD-L1^{+} (R4) and CD11b^{+}/CD33^{+} (R5). To reduce the use of fluorophores and following a previous group study (Rico, Aguilar Hernández, et al., 2021), PMN-MDSC and M-MDSC are classified based on their scatter properties and CD33/CD11b expression, being the polymorphonuclear CD33^{high}/CD11b^{+} and the monocytic CD11b^{high}/CD33^{+}.

For the bioinformatic analysis, FCS are ordered according to stimulation group (DMSO or PMA) and then concatenated, resulting in one large single file of about 68 million events. In this case, a downsample step is required to run the t-SNE (T-distributed Stochastic Neighbor Embedding) and UMAP (Uniform Manifold Approximation and Projection) algorithms. For t-SNE calculation, the concatenation is programmed to include up to 2 million cells equally taken from each sample. The 2-dimension t-SNE reduction algorithm is performed selecting the variables that correspond to the pulse height of the following markers: CD33, CD11b, HLA-DR and PD-L1. Default settings are applied: 30% perplexity and 1000 iterations. UMAP is run for CD33-H, CD11b-H, HLA-DR-H and PD-L1-H markers from a downsampled file of $5x10^{5}$ cells, with the default parameters of 15 nearest neighbors, 0.5 of minimum distance and 2 components. The Flow Self-Organizing Map (FlowSOM) is performed on the whole concatenated file and selecting the same four markers (CD33-H, CD11b-H, HLA-DR-H and PD-L1-H). The parameter scale is applied to normalize each marker expression, as well as the automatic number of meta clusters to avoid losing clusters by setting a limiting number. The analysis of DMSO and PMA subgroups is then performed based on the previously generated global structure.

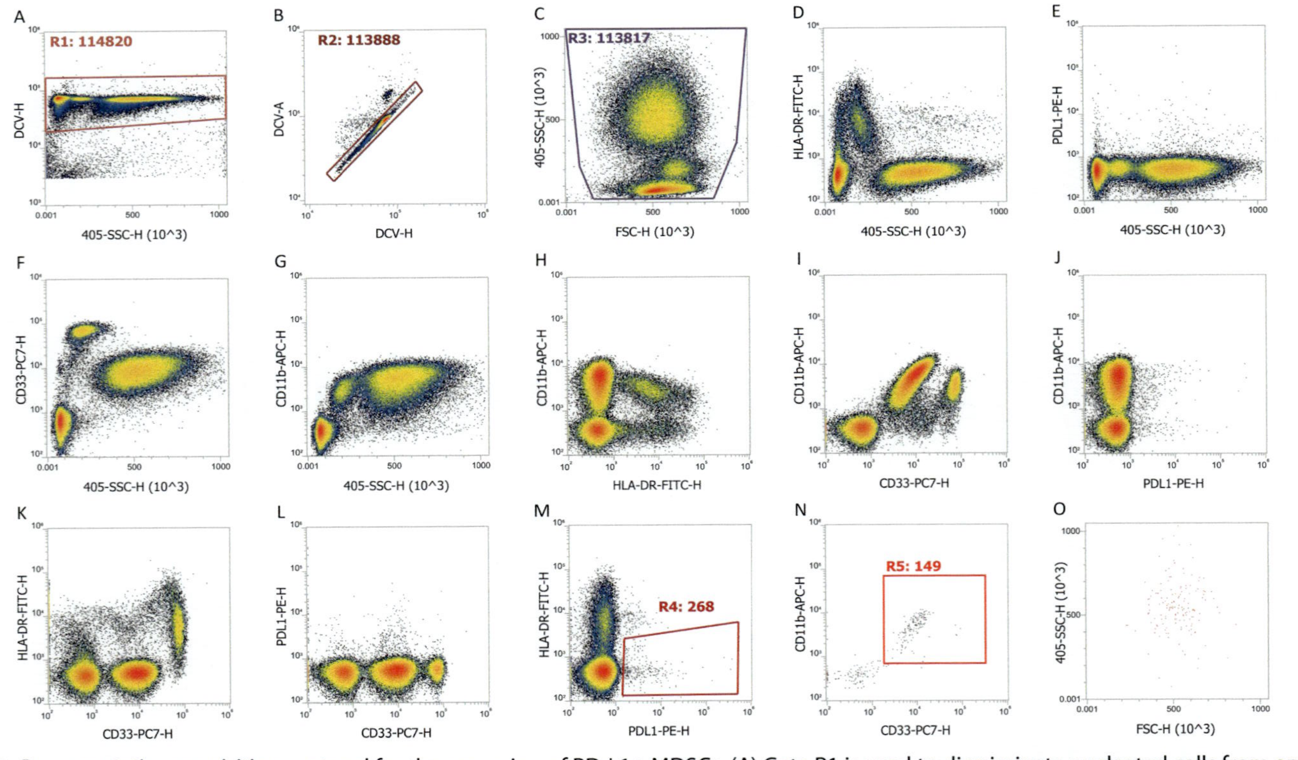

Fig. 1 Representative acquisition protocol for the screening of PD-L1+ MDSCs. (A) Gate R1 is used to discriminate nucleated cells from erythrocytes and debris. (B) R2 is used to discriminate doublets using DNA staining. (C) R3 is used to finally discriminate apoptotic cells and debris. (D–G) represent FITC-HLA (D), PE-PD-L1 (E), PE-Cy7-CD33 (F), and APC-CD11b (G) versus SSC. (H–L) represent the combination of all fluorochromes to adjust the color compensation. (M–O) represent the PD-L1 + cells gating strategy. These cells were selected as HLA-DR$^{low/neg}$ / PD-L1^{+} (R4) and CD11b^{+}/CD33^{+} (R5) population. Samples were acquired using Attune™ NxT Flow Cytometer (Thermo Fisher).

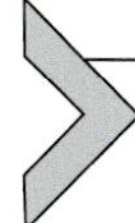

3. Representative results

3.1 PD-L1 shows a differential expression pattern in MDSCs after stimulation

The PD-L1 reactivity of MDSCs was determined in bone marrow of 35 MM patients with and without PMA stimulation. Upon stimulation, PD-L1 expression increased in 94.3% of patients, ranging from 1 to 650 times higher compared to unstimulated samples (Fig. 2A). These patients that respond to stimulation indicate that PD-L1 is already expressed, and PMA stimulation allows the recognition of the molecule. Interestingly, some patients do not respond to PMA stimulation and display the same PD-L1 expression as without stimulation (Fig. 2B). These patients, named non-responders in regard to PMA stimulation, can be identified with this minimal sample perturbation protocol. Additionally, a kinetic analysis was performed to determine the optimal stimulation time of PMA and, therefore, to get the maximum PD-L1 reactivity. The higher recognition of PD-L1 was recorded at 5 min, after which PD-L1 levels started decreasing (Fig. 2C).

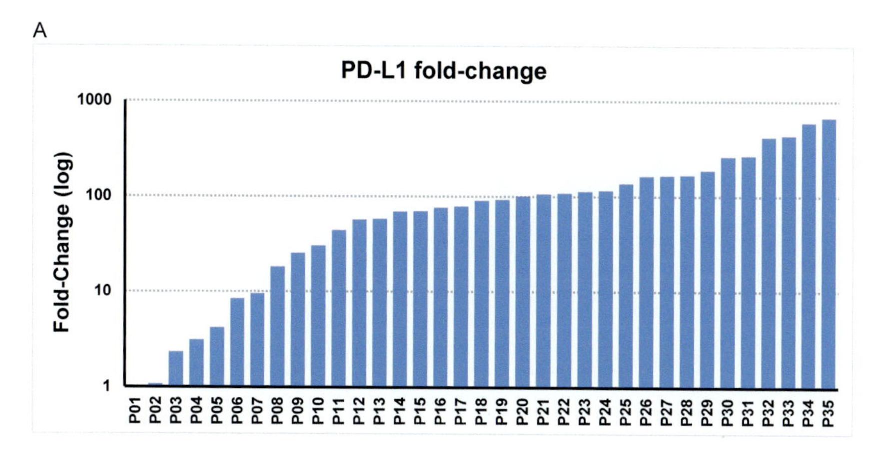

Fig. 2 PD-L1 fold-change in a series of 35 multiple myeloma patients. (A) Fold-change is calculated as ratio of number of stimulated and non-stimulated MDSCs PD-L1$^+$. Patients P01 to P35 showed a wide fold-change variation, ranging from 1 (no variation) to 650. (B) Representative cases of a non-responding patient (P02) with PD-L1 fold-change ≤ 1 (upper row) and a responding patient (P26) with a PD-L1 fold-change >1 (lower row). (C) Additionally, PD-L1 was determined after 1 to 60 min of PMA stimulation. PD-L1$^+$ cells are represented in terms of cell counts.

(Continued)

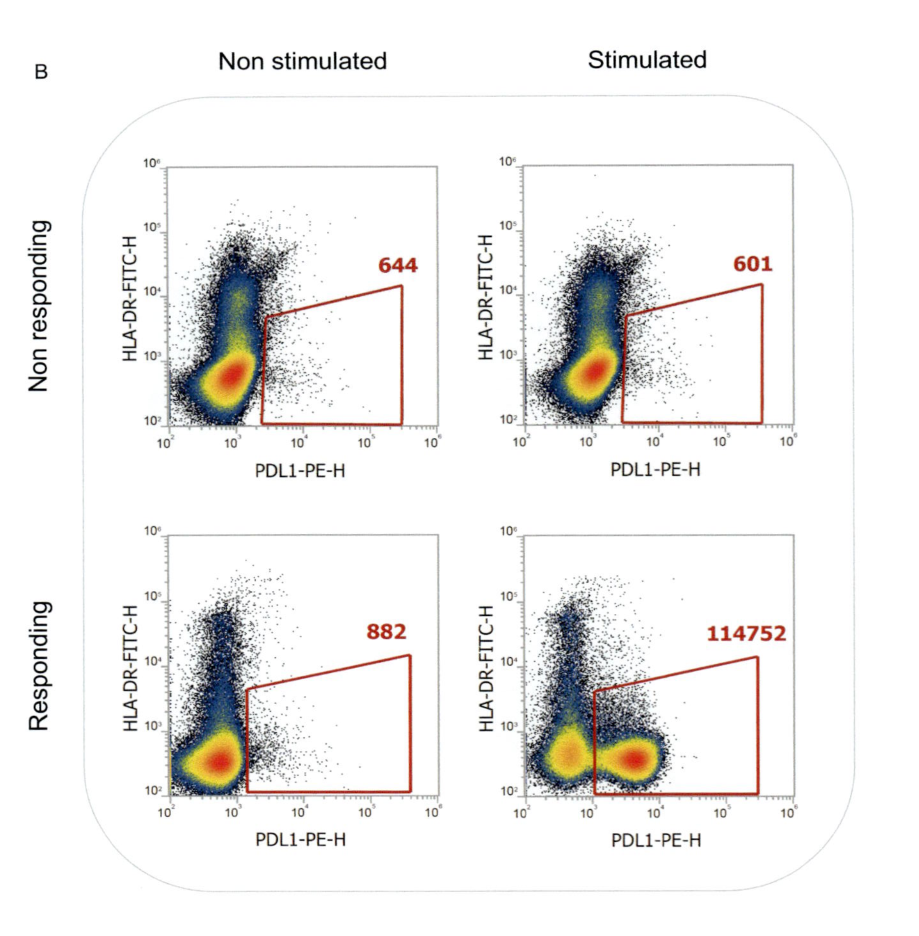
B
Non stimulated
Stimulated
Non responding
Responding
HLA-DR-FITC-H
PDL1-PE-H
644
601
882
114752

Fig. 2—Cont'd

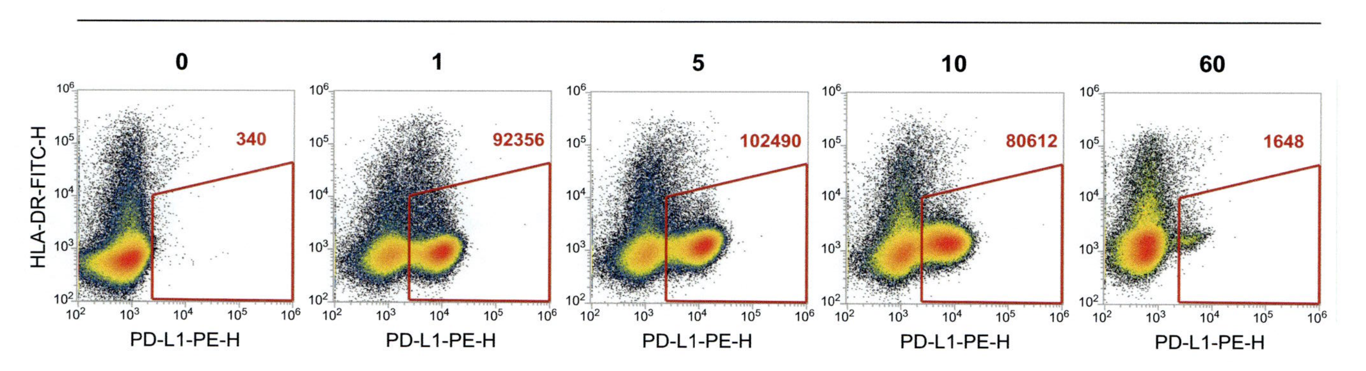
C
PMA stimulation time (minutes)
0
1
5
10
60
HLA-DR-FITC-H
PD-L1-PE-H
340
92356
102490
80612
1648

Fig. 2—Cont'd

3.2 PD-L1 is not expressed at cytoplasmatic level

Based on the PD-L1 fold-change from stimulated to unstimulated samples, we hypothesize PD-L1 could be expressed in the cytoplasm and translocated to the cell membrane upon stimulation with PMA. To confirm that, we developed a protocol to study PD-L1 cytoplasmatic levels. An additional marker, CD11b, was studied as a positive control of cytoplasmatic expression. The cytoplasmatic expressions of PD-L1 and CD11b were simultaneously studied in 11 patients and, whereas CD11b cytoplasmatic levels were reported, no PD-L1 protein could be detected in the cytoplasm (Fig. 3). Therefore, PD-L1 is not translocated to the membrane but expressed there. Contrarily, CD11b is expressed in the cytoplasm and translocated to the cell surface upon stimulation.

3.3 PD-L1 monoclonal antibody and Durvalumab are competing for the same PD-L1 binding site

To make sure that the anti-PD-L1 PE-conjugated monoclonal antibody recognizes the same binding site of the target molecule as the immunotherapy

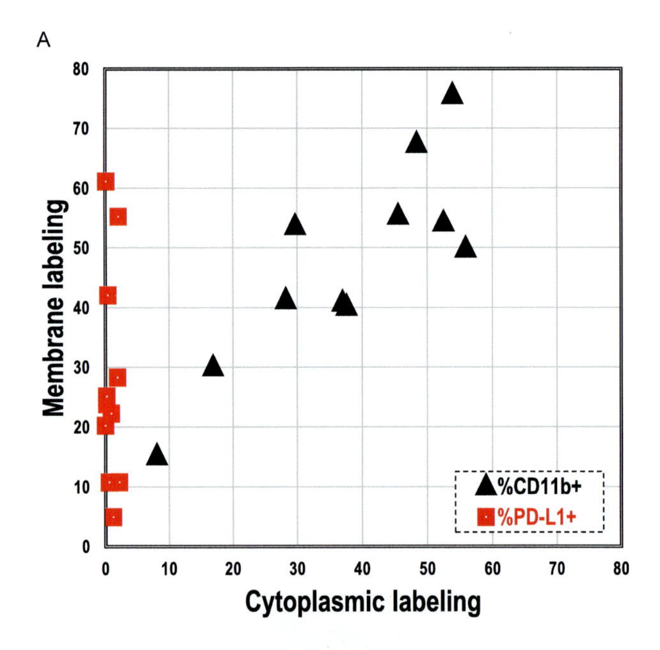

Fig. 3 Determination of PD-L1 and CD11b cell surface and cytoplasmic levels. (A) Comparison of %CD11b+and %PD-L1+ in cell surface and cytoplasm in a series of 11 multiple myeloma patients. (B) Results of %CD11b+and %PD-L1+ obtained in cytoplasm and membrane (basal and stimulated). (C) Representative case (P30) of PD-L1 and CD11b cell surface determination: non-stimulated and stimulated and cytoplasmic determination.

(Continued)

B

	Cytoplasm		Membrane			
	%CD11b+	%PD-L1+	% CD11b+		% PD-L1+	
			Non stimulated	Stimulated	Non stimulated	Stimulated
P04	37.62	0.61	2.88	40.15	3.87	10.84
P07	8.11	1.24	3.85	15.16	0.76	4.98
P10	55.87	2.23	13.77	49.83	0.60	10.87
P18	28.21	0.22	0.74	41.29	1.06	23.89
P19	53.85	0.16	73.20	75.55	1.73	61.19
P20	29.70	1.95	50.29	53.65	0.87	28.35
P21	52.5	0.95	8.97	54.23	0.83	22.32
P23	36.98	0.03	17.62	40.86	0.61	20.32
P24	48.36	2.09	40.86	67.40	1.06	55.31
P25	45.52	0.43	33.56	55.37	1.42	42.13
P30	16.90	0.23	9.20	30.02	3.63	25.19

C

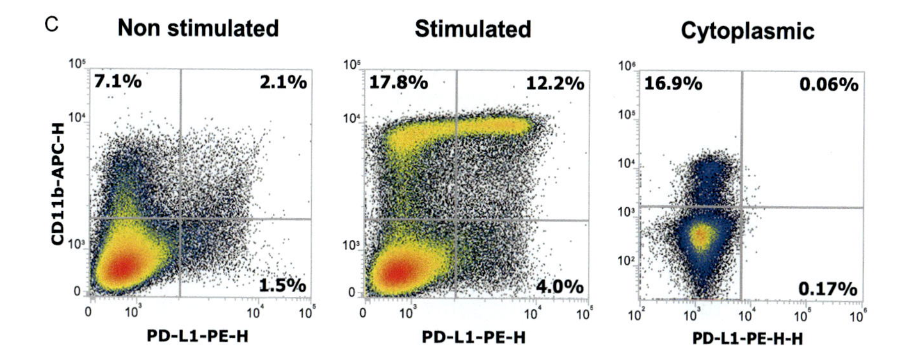

Fig. 3—Cont'd

antibody, a competition assay with Durvalumab, an anti-PD-L1 immunotherapy drug, is performed.

Here we show representative results from four bone marrow from multiple myeloma patients (P36 to P39) prepared for the competition assay (Fig. 4). The recognition of PD-L1 was decreased with higher concentrations of Durvalumab, meaning that PE-PD-L1 and Durvalumab compete for the same binding site. In the case of P38, the binding of PE-PD-L1 declined a bit but not as sharply as the former ones. Surprisingly, Durvalumab could not bind to PD-L1 of the P37 patient, and the recognition of the ligand was entirely carried out by the monoclonal antibody. Unfortunately, the anti-PD-L1 reactivity and binding characteristics of Durvalumab have not yet been elucidated (Kumar et al., 2019; Tan et al., 2018) and we cannot compare the binding sites of the two antibodies. We hypothesize that P37 and P38 have some mutations that partly modify the aminoacidic composition and structure of the PD-L1 molecule, hence changing the Durvalumab affinity.

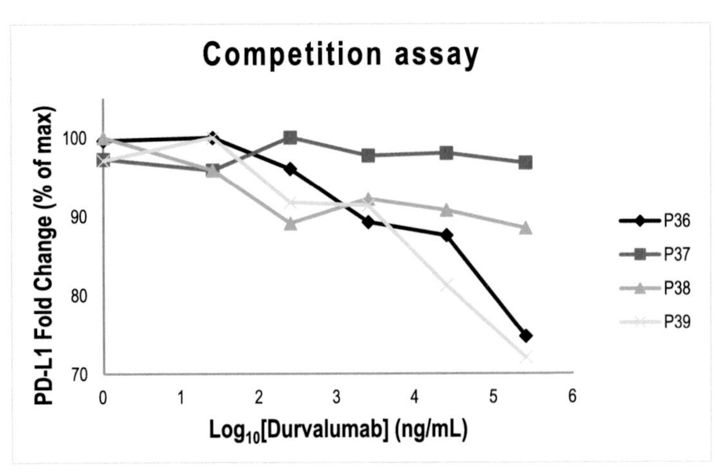

Fig. 4 Competition experiments between Durvalumab (immunotherapy drug) and PE-PD-L1 (monoclonal antibody). Cells were stimulated and incubated with 2.5 ng/μL PE-PD-L1 and increasing concentrations of Durvalumab (0, 0.025, 0.25, 2.5, 25 and 250 ng/μL). PE-PD-L1 expression levels were compared with non-stimulated cells.

3.4 Dimensionality reduction and visualization of the obtained data

The 2-dimension reduction algorithm t-SNE distributes the data retaining the local structure while displaying some global structure such as clusters at multiple scales. t-SNE determines the local neighborhood size for each datapoint separately based on the local density of the data (Van Der Maaten & Hinton, 2008). t-SNE is currently the most commonly used technique in single-cell analysis. Similarly to t-SNE, the Uniform Manifold Approximation and Projection (UMAP) is a nonlinear dimensionality-reduction technique to analyze high-dimensional data. Compared to t-SNE, UMAP better preserves large-scale distances with superior run time performance (Becht et al., 2019; Mcinnes, Healy, & Melville, 2020). Considering another technique of data analysis, we performed FlowSOM, an algorithm that builds self-organizing maps using a two-level clustering and star charts, in order to display clusters of marker expression on all cells and therefore reveal cell subsets (Van Gassen et al., 2015).

All three mentioned approximations allow the visualization of the data from different perspectives. As shown in Fig. 5, DMSO and PMA do not have a full complementary distribution, meaning that PD-L1$^+$ cells are not restricted to PMA-stimulated samples and the other way around, that some PMA samples do not display PD-L1$^+$ cells (non-responders,

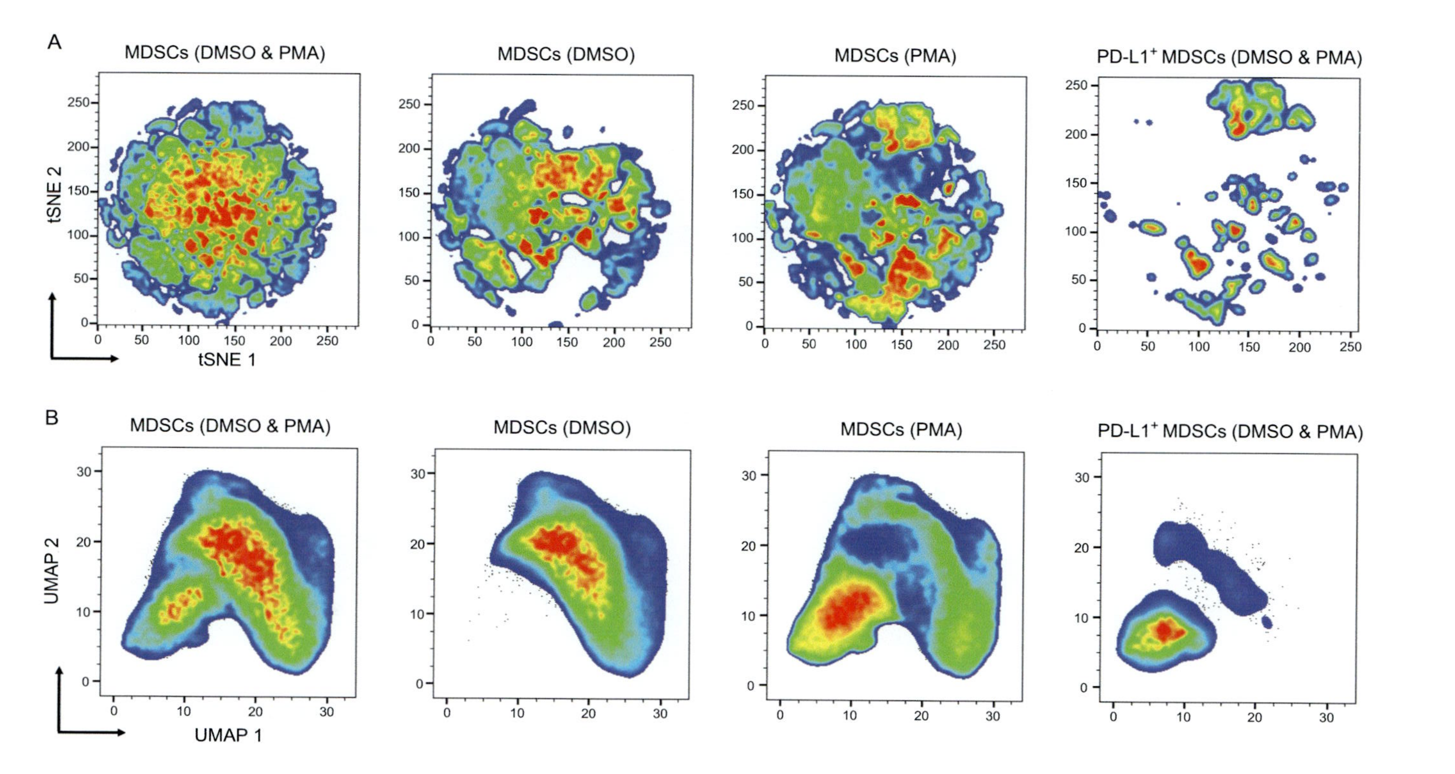

Fig. 5 The bioinformatic algorithms t-SNE, UMAP and FlowSOM grant a global and broad visualization of the data. Differences in PD-L1$^+$ detection upon DMSO and PMA stimulation are shown in t-SNE (A), UMAP (B, C) and FlowSOM (D) read-outs. C. UMAP contour plots of PD-L1$^+$ MDSCs of the global data (in red) against MDSCs from DMSO or PMA stimulated subsets (in black).

(Continued)

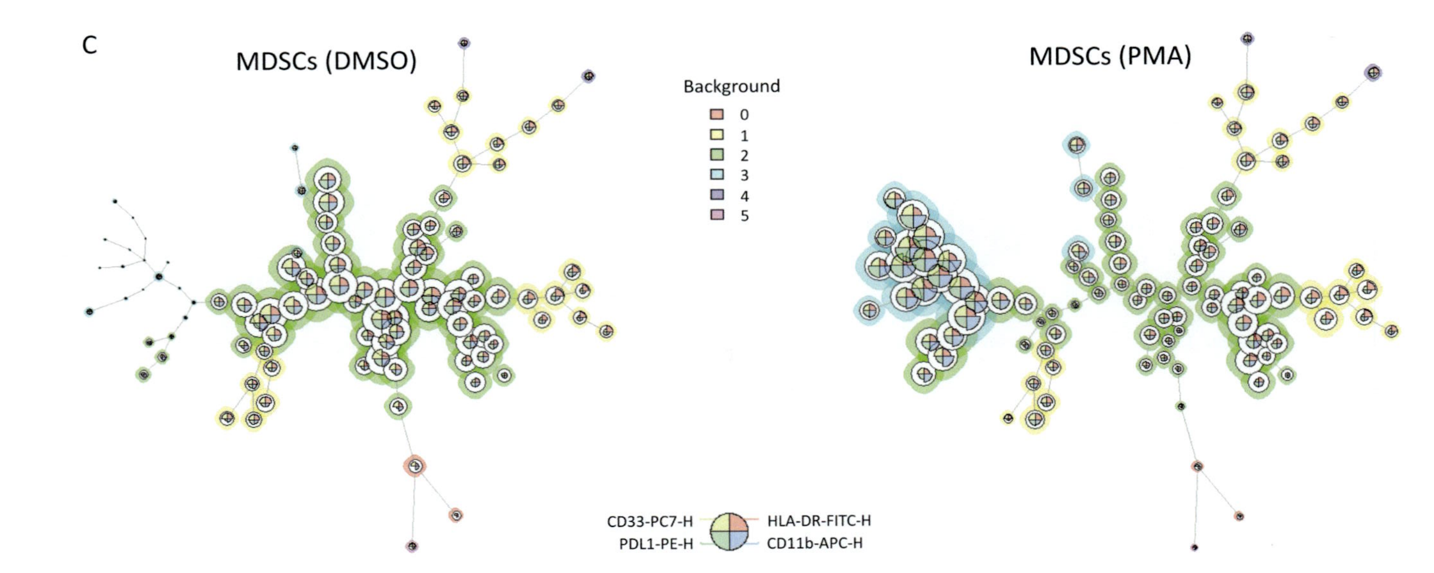
C
MDSCs (DMSO)
MDSCs (PMA)
Background
0
1
2
3
4
5
CD33-PC7-H
PDL1-PE-H
HLA-DR-FITC-H
CD11b-APC-H

Fig. 5—Cont'd

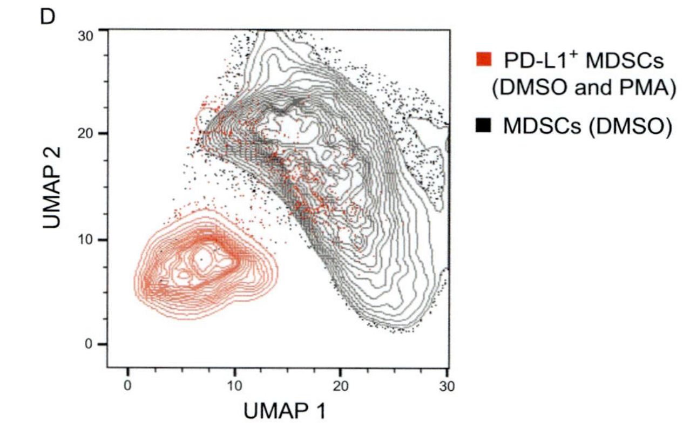
D
UMAP 2
UMAP 1
0
10
20
30
PD-L1+ MDSCs (DMSO and PMA)
MDSCs (DMSO)

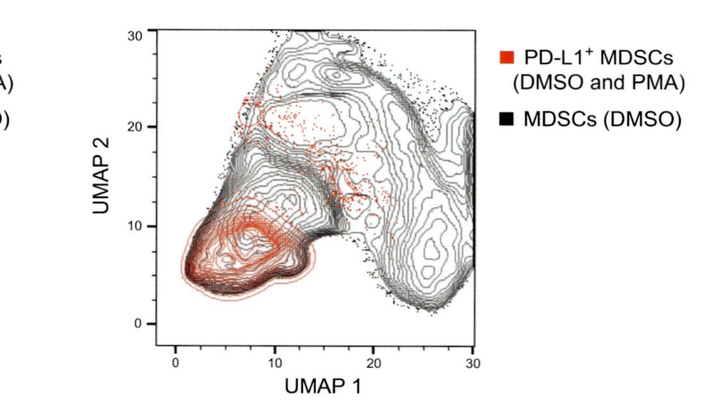
UMAP 2
UMAP 1
0
10
20
30
PD-L1+ MDSCs (DMSO and PMA)
MDSCs (DMSO)

Fig. 5—Cont'd

see Fig. 2B). Indeed, the unstimulated group has patients with PD-L1$^+$ cells, indicating that some patients have a group of MDSCs with a PD-L1 basal conformation similar to the stimulated conformation and, in consequence, a higher basal PD-L1 recognition.

4. Discussion

Little is known about how the variability of PD-L1 affects the response to immunotherapy. Nevertheless, this heterogeneity is likely to contribute to the therapy ineffectiveness in some patients.

In this chapter, we show a methodology to predict PD-L1 immunotherapy success in multiple myeloma by using bone marrow samples. Screening PD-L1 in MDSCs is a rapid method and a representative study of the ligand status. This methodology can also be applied to study other hematological diseases as well as solid tumors and with other immunotherapies than the PD-1/PD-L1 immune checkpoint (Rico, Aguilar Hernández, et al., 2021). For assessing PD-L1 reactivity, it is important to avoid any functional or conformational change in the molecule. In this fragile context, the processing of the sample must be minimal to better preserve the native structure of the molecule. In addition, a lysing procedure would deplete part of the myeloid-derived suppressor cell population, since PMN-MDSC have a neutrophil-like phenotype (Perez et al., 2020) and neutrophils are particularly sensitive to lysis. Therefore, the samples are not lysed nor washed, discarding erythrocytes and platelets by setting a fluorescence threshold on nucleated cells (Petriz et al., 2018; Rico, Salvia, et al., 2021). Overall, no-lyse no-wash methodologies in combination with functional assays show promise as an emerging strategy to model conformational changes or structure in the target site.

The screening assay has allowed us to identify different basal and stimulated PD-L1 conformation patterns among a group of multiple myeloma patients, that can be related with immunotherapy success. Also, the distinction of the two subtypes of MDSCs can correlate to prognosis, PMN-MDSC being a bad indicator (Görgün et al., 2013). Based on our results, the PMA stimulation apparently enables a conformational change which dramatically increases the union ligand-antibody. Despite this, there still are some MM patients, the non-responders, that do not express or unfold PD-L1 upon stimulation. Moreover, competition assays have enabled us to identify patients with PD-L1 structural mutations that avoid the union of Durvalumab and, consequently, the ineffectiveness of the immunotherapy.

The PMA compound has pleiotropic effects and its action upon PD-L1 is unknown. Being described the enhanced accessibility of PD-L1 after deglycosylation (Lee, Wang et al., 2019), one logic hypothesis would be that PMA deglycosylates PD-L1, increasing its recognition. Nevertheless, the PMA kinetics assay (Fig. 3) clearly shows that after reaching the point of higher reactivity to antibodies, PD-L1 lowers its accessibility back to unstimulated values. This behavior cannot be explained by a reglycosylation, since glycosylation is typically performed in the cytoplasm instead of on the membrane surface and it typically requires more time (Hevér, Darula, & Medzihradszky, 2019). According to the conformational transition model proposed by Ghiotto et al. (2010), we hypothesize that the PMA compound has a role in modifying the conformational disposition of PD-L1 to a more accessible one, which could be similar to that bound to PD-1.

For the improvement of multiple myeloma and other solid tumors treatment, it is essential to obtain a new immunotherapy that is efficient knocking down the PD-1/PD-L1 checkpoint in those patients that are refractory to the current immunotherapies. Finding a non-toxic agent that can modulate the PD-L1 accessibility promises a great potential as immunotherapy coadjuvant, since this putative factor could enable therapies to successfully interact with the ligand and overcome drug resistance. A deglycosylating agent could improve the ligand accessibility but it would deregulate other biological pathways. On the contrary, a molecule or compound able to reproduce the transitory and unfolding effect of PMA would be a suitable candidate for immunotherapy coadjuvant.

Little information about PD-L1 structure dynamics and recognition can be found in the literature, and more studies are needed to analyze the immunophenotypic characteristics of each patient and link them to clinic outcome. Flow cytometry stands as an encouraging technique to permit the classification of cancer patients based on their PD-L1 levels for the administration of the most appropriate and personalized treatment. Chimeric antigen receptor (CAR) T-cell therapy may also be a new potential field of study, not also based on targeted antigen downregulation, to look at possible conformational changes.

Critical assessment of PD-L1 folding, as well as those targets having similar unexpected features, may help to develop better treatment strategies or to predict therapy resistance.

Acknowledgments

This study was partially supported by the Josep Carreras Foundation and by Thermo Fisher Scientific. We thank the CERCA Programme/Generalitat de Catalunya, the Josep Carreras Foundation and Germans Trias i Pujol Research Institute for institutional support, and we acknowledge financial support from the Obra Social la Caixa. The authors are very grateful to Clara Streiff, Paola Paglia, Sergio Ramon, Lluís Sainz, and Víctor Querol from Thermo Fisher Scientific for all their help in this research field, and to Sara Vergara and Minerva Raya for kindly providing samples used in this study.

Disclosures

M.D.W. and J.A.B. are employees of Thermo Fisher Scientific, which is in the business of selling flow cytometers and flow cytometry reagents. The rest of the authors declares no competing interests.

References

Ahmed, M., & Barakat, K. (2017). The too many faces of PD-L1: A comprehensive conformational analysis study. *Biochemistry, 56*(40), 5428–5439. https://doi.org/10.1021/acs.biochem.7b00655.

Badros, A., Hyjek, E., Ma, N., Lesokhin, A., Dogan, A., Rapoport, A. P., et al. (2017). Pembrolizumab, pomalidomide, and low-dose dexamethasone for relapsed/refractory multiple myeloma. *Blood, 130*(10), 1189–1197. https://doi.org/10.1182/BLOOD-2017-03-775122.

Bar, N., Costa, F., Das, R., Duffy, A., Samur, M., McCachren, S., et al. (2020). Differential effects of PD-L1 versus PD-1 blockade on myeloid inflammation in human cancer. *JCI Insight, 5*(12). https://doi.org/10.1172/jci.insight.129353.

Becht, E., McInnes, L., Healy, J., Dutertre, C. A., Kwok, I. W. H., Ng, L. G., et al. (2019). Dimensionality reduction for visualizing single-cell data using UMAP. *Nature Biotechnology, 37*(1), 38–47. https://doi.org/10.1038/nbt.4314.

Binsfeld, M., Muller, J., Lamour, V., De Veirman, K., De Raeve, H., Bellahcène, A., et al. (2016). Granulocytic myeloid-derived suppressor cells promote angiogenesis in the context of multiple myeloma. *Oncotarget*, 7(25), 37931–37943. https://doi.org/10.18632/oncotarget.9270.

Brigle, K., & Rogers, B. (2017). Pathobiology and diagnosis of multiple myeloma. *Seminars in Oncology Nursing, 33*(3), 225–236. https://doi.org/10.1016/j.soncn.2017.05.012.

D'Souza, A., Hari, P., Pasquini, M., Braun, T., Johnson, B., Lundy, S., et al. (2019). A phase 2 study of Pembrolizumab during Lymphodepletion after autologous hematopoietic cell transplantation for multiple myeloma. *Biology of Blood and Marrow Transplantation, 25*(8), 1492–1497. https://doi.org/10.1016/J.BBMT.2019.04.005.

Dhodapkar, M. V., Sexton, R., Das, R., Dhodapkar, K. M., Zhang, L., Sundaram, R., et al. (2015). Prospective analysis of antigen-specific immunity, stem-cell antigens, and immune checkpoints in monoclonal gammopathy. *Blood, 126*(22), 2475–2478. https://doi.org/10.1182/blood-2015-03-632919.

Dysthe, M., & Parihar, R. (2020). Myeloid-derived suppressor cells in the tumor microenvironment. *Advances in Experimental Medicine and Biology, 1224*, 117–140. https://doi.org/10.1007/978-3-030-35723-8_8.

Filipazzi, P., Valenti, R., Huber, V., Pilla, L., Canese, P., Iero, M., et al. (2007). Identification of a new subset of myeloid suppressor cells in peripheral blood of melanoma patients with modulation by a granulocyte-macrophage colony-stimulation factor-based antitumor vaccine. *Journal of Clinical Oncology, 25*(18), 2546–2553. https://doi.org/10.1200/JCO.2006.08.5829.

Freeman, B. G. J., Long, A. J., Iwai, Y., Bourque, K., Chernova, T., Nishimura, H., et al. (2000). Engagement of the PD-1 Immunoinhibitory receptor by a novel B7 family member leads to negative regulation of lymphocyte activation. *Journal of Experimental Medicine, 192*(7), 1028–1034.

Gabrilovich, D. I., & Nagaraj, S. (2009). Myeloid-derived-suppressor cells as regulators of the immune system. *Nature Reviews. Immunology*, *9*(3), 162. https://doi.org/10.1038/NRI2506.

Ghiotto, M., Gauthier, L., Serriari, N., Pastor, S., Truneh, A., Nunès, J. A., et al. (2010). PD-L1 and PD-L2 differ in their molecular mechanisms of interaction with PD-1. *International Immunology*, *22*(8), 651–660. https://doi.org/10.1093/intimm/dxq049.

Görgün, G. T., Whitehill, G., Anderson, J. L., Hideshima, T., Maguire, C., Laubach, J., et al. (2013). Tumor-promoting immune-suppressive myeloid-derived suppressor cells in the multiple myeloma microenvironment in humans. *Blood*, *121*(15), 2975–2987. https://doi.org/10.1182/blood-2012-08-448548.

Hevér, H., Darula, Z., & Medzihradszky, K. F. (2019). Characterization of site-specific N-glycosylation. *Methods in Molecular Biology*, *1934*, 93–125. https://doi.org/10.1007/978-1-4939-9055-9_8.

Ishida, Y., Agata, Y., Shibahara, K., & Honjo, T. (1992). Induced expression of PD-1, a novel member of the immunoglobulin gene superfamily, upon programmed cell death. *EMBO Journal*, *11*(11), 3887–3895. https://doi.org/10.1002/j.1460-2075.1992.tb05481.x.

Ishida, M., Iwai, Y., Tanaka, Y., Okazaki, T., Freeman, G. J., Minato, N., et al. (2002). Differential expression of PD-L1 and PD-L2, ligands for an inhibitory receptor PD-1, in the cells of lymphohematopoietic tissues. *Immunology Letters*, *84*(1), 57–62. https://doi.org/10.1016/S0165-2478(02)00142-6.

Iwata, M., Ohoka, Y., Kuwata, T., & Asada, A. (1996). Regulation of T cell apoptosis via T cell receptors and steroid receptors. *Stem Cells*, *14*(6), 632–641. https://doi.org/10.1002/stem.140632.

Kelly, K. R., Espitia, C. M., Zhao, W., Wu, K., Visconte, V., Anwer, F., et al. (2018). Oncolytic reovirus sensitizes multiple myeloma cells to anti-PD-L1 therapy. *Leukemia*, *32*(1), 230–233. https://doi.org/10.1038/leu.2017.272.

Kumar, D., Lisok, A., Dahmane, E., McCoy, M., Shelake, S., Chatterjee, S., et al. (2019). Peptide-based PET quantifies target engagement of PD-L1 therapeutics. *Journal of Clinical Investigation*, *129*(2), 616–630. https://doi.org/10.1172/JCI122216.

Kumar, S. K., Rajkumar, V., Kyle, R. A., Van Duin, M., Sonneveld, P., Mateos, M. V., et al. (2017). Multiple myeloma. *Nature Reviews Disease Primers*, *3*, 1–20. https://doi.org/10.1038/nrdp.2017.46.

Kwon, M., Kim, C. G., Lee, H., Cho, H., Kim, Y., Lee, E. C., et al. (2020). PD-1 blockade reinvigorates bone marrow CD8+ T cells from patients with multiple myeloma in the presence of TGFb inhibitors. *Clinical Cancer Research*, *26*(7), 1644–1655. https://doi.org/10.1158/1078-0432.CCR-19-0267.

Latchman, Y., Wood, C. R., Chernova, T., Chaudhary, D., Borde, M., Chernova, I., et al. (2001). PD-L2 is a second ligand for PD-1 and inhibits T cell activation. *Nature Immunology*, *2*(3), 261–268. https://doi.org/10.1038/85330.

Lee, H. T., Lee, S. H., & Heo, Y.-S. (2019). Molecular interactions of antibody drugs targeting PD-1, PD-L1, and CTLA-4 in immuno-oncology. *Molecules*, *24*(6). https://doi.org/10.3390/molecules24061190.

Lee, H. H., Wang, Y. N., Xia, W., Chen, C. H., Rau, K. M., Ye, L., et al. (2019). Removal of N-linked glycosylation enhances PD-L1 detection and predicts anti-PD-1/PD-L1 therapeutic efficacy. *Cancer Cell*, *36*(2), 168–178.e4. https://doi.org/10.1016/j.ccell.2019.06.008.

Li, C. W., Lim, S. O., Chung, E. M., Kim, Y. S., Park, A. H., Yao, J., et al. (2018). Eradication of triple-negative breast Cancer cells by targeting glycosylated PD-L1. *Cancer Cell*, *33*(2), 187–201.e10. https://doi.org/10.1016/j.ccell.2018.01.009.

Li, C. W., Lim, S. O., Xia, W., Lee, H. H., Chan, L. C., Kuo, C. W., et al. (2016). Glycosylation and stabilization of programmed death ligand-1 suppresses T-cell activity. *Nature Communications*, 7. https://doi.org/10.1038/ncomms12632.

Mateos, M. V., Orlowski, R. Z., Ocio, E. M., Rodríguez-Otero, P., Reece, D., Moreau, P., et al. (2019). Pembrolizumab combined with lenalidomide and low-dose dexamethasone for relapsed or refractory multiple myeloma: Phase I KEYNOTE-023 study. *British Journal of Haematology*, *186*(5), e117–e121. Blackwell Publishing Ltd https://doi.org/10.1111/bjh.15946.

Mcinnes, L., Healy, J., & Melville, J. (2020). *UMAP: Uniform manifold approximation and projection for dimension reduction*. https://doi.org/10.48550/arXiv.1802.03426. ArXiv, abs/1802.03426.

Minnie, S. A., & Hill, G. R. (2020). Immunotherapy of multiple myeloma. *Journal of Clinical Investigation*, *130*(4), 1565–1575. https://doi.org/10.1172/JCI129205.

Paiva, B., Azpilikueta, A., Puig, N., Ocio, E. M., Sharma, R., Oyajobi, B. O., et al. (2015). PD-L1/PD-1 presence in the tumor microenvironment and activity of PD-1 blockade in multiple myeloma. *Leukemia*, *29*(10), 2110–2113. https://doi.org/10.1038/leu.2015.79.

Perez, C., Botta, C., Zabaleta, A., Puig, N., Cedena, M.-T., Goicoechea, I., et al. (2020). Immunogenomic identification and characterization of granulocytic myeloid-derived suppressor cells in multiple myeloma. *Blood*, *136*(2), 199–209.

Petriz, J., Bradford, J. A., & Ward, M. D. (2018). No lyse no wash flow cytometry for maximizing minimal sample preparation. *Methods*, *134–135*, 149–163. https://doi.org/10.1016/j.ymeth.2017.12.012.

Ramachandran, I., Martner, A., Pisklakova, A., Condamine, T., Chase, T., Vogl, T., et al. (2013). Myeloid derived suppressor cells regulate growth of multiple myeloma by inhibiting T cells in bone marrow. *Journal of Immunology*, *190*(7), 3815–3823. https://doi.org/10.4049/jimmunol.1203373.

Rico, L. G., Aguilar Hernández, A., Ward, M. D., Bradford, J. A., Juncà, J., Rosell, R., et al. (2021). Unmasking the expression of PD-L1 in myeloid derived suppressor cells: A case study in lung cancer to discover new drugs with specific on-target efficacy. *Translational Oncology*, *14*(1), 2020–2022. https://doi.org/10.1016/j.tranon.2020.100969.

Rico, L. G., Salvia, R., Ward, M. D., Bradford, J. A., & Petriz, J. (2021). Flow-cytometry-based protocols for human blood/marrow immunophenotyping with minimal sample perturbation. *STAR Protocols*, *2*(4). https://doi.org/10.1016/j.xpro.2021.100883.

Siegel, R. L., Miller, K. D., Fuchs, H. E., & Jemal, A. (2021). Cancer statistics, 2021. *CA: A Cancer Journal for Clinicians*, *71*(1), 7–33. https://doi.org/10.3322/caac.21654.

Soekojo, C. Y., Ooi, M., de Mel, S., & Chng, W. J. (2020). Immunotherapy in multiple myeloma. *Cells*, *9*(3). https://doi.org/10.3390/cells9030601.

Tamura, H., Dan, K., Tamada, K., Nakamura, K., Shioi, Y., Hyodo, H., et al. (2005). Expression of functional B7-H2 and B7.2 costimulatory molecules and their prognostic implications in de novo acute myeloid leukemia. *Clinical Cancer Research*, *11*(16), 5708–5717. https://doi.org/10.1158/1078-0432.CCR-04-2672.

Tan, S., Liu, K., Chai, Y., Zhang, C. W. H., Gao, S., Gao, G. F., et al. (2018). Distinct PD-L1 binding characteristics of therapeutic monoclonal antibody durvalumab. *Protein & Cell*, *9*(1), 135–139. https://doi.org/10.1007/s13238-017-0412-8.

Tseng, S. Y., Otsuji, M., Gorski, K., Huang, X., Slansky, J. E., Pai, S. I., et al. (2001). B7-DC, a new dendritic cell molecule with potent costimulatory properties for T cells. *Journal of Experimental Medicine*, *193*(7), 839–845. https://doi.org/10.1084/jem.193.7.839.

Van Der Maaten, L., & Hinton, G. (2008). Visualizing Data using t-SNE. *Journal of Machine Learning Research*, *9*, 2579–2605.

Van Gassen, S., Callebaut, B., Van Helden, M. J., Lambrecht, B. N., Demeester, P., Dhaene, T., et al. (2015). FlowSOM: Using self-organizing maps for visualization and interpretation of cytometry data. *Cytometry. Part A: The Journal of the International Society for Analytical Cytology*, *87*(7), 636–645. https://doi.org/10.1002/CYTO.A.22625.

Van Valckenborgh, E., Schouppe, E., Movahedi, K., De Bruyne, E., Menu, E., De Baetselier, P., et al. (2012). Multiple myeloma induces the immunosuppressive capacity of distinct myeloid-derived suppressor cell subpopulations in the bone marrow. *Leukemia, 26*, 2424–2428. https://doi.org/10.1038/leu.2012.113.

Wang, J., Veirman, K. D., Beule, N. D., Maes, K., Bruyne, E. D., Valckenborgh, E. V., et al. (2015). The bone marrow microenvironment enhances multiple myeloma progression by exosome-mediated activation of myeloid-derived suppressor cells. *Oncotarget, 6*(41), 43992–44004. https://doi.org/10.18632/oncotarget.6083.

Wang, Z., Zhang, L., Wang, H., Xiong, S., Li, Y., Tao, Q., et al. (2015). Tumor-induced CD14+HLA-DR−/low myeloid-derived suppressor cells correlate with tumor progression and outcome of therapy in multiple myeloma patients. *Cancer Immunology, Immunotherapy, 64*(3), 389–399. https://doi.org/10.1007/s00262-014-1646-4.

Xue, J., Chen, C., Qi, M., Huang, Y., Wang, L., Gao, Y., et al. (2017). Type Iγ phosphatidylinositol phosphate kinase regulates PD-L1 expression by activating NF-κB. *Oncotarget, 8*(26), 42414–42427. https://doi.org/10.18632/oncotarget.17123.

Yamazaki, T., Akiba, H., Iwai, H., Matsuda, H., Aoki, M., Tanno, Y., et al. (2002). Expression of programmed death 1 ligands by murine T cells and APC. *The Journal of Immunology, 169*(10), 5538–5545. https://doi.org/10.4049/jimmunol.169.10.5538.

Yang, Y., Li, C. W., Chan, L. C., Wei, Y., Hsu, J. M., Xia, W., et al. (2018). Exosomal PD-L1 harbors active defense function to suppress t cell killing of breast cancer cells and promote tumor growth. *Cell Research, 28*(8), 862–864. https://doi.org/10.1038/s41422-018-0060-4.

Yin, Z., Yu, M., Ma, T., Zhang, C., Huang, S., Karimzadeh, M. R., et al. (2021). Mechanisms underlying low-clinical responses to PD-1/PD-L1 blocking antibodies in immunotherapy of cancer: A key role of exosomal PD-L1. *Journal for Immunotherapy of Cancer, 9*(1), 1–9. https://doi.org/10.1136/jitc-2020-001698.

Zafar Usmani, S., Schjesvold, F., Oriol, A., Karlin, L., Cavo, M., Rifkin, R. M., et al. (2019). Pembrolizumab plus lenalidomide and dexamethasone for patients with treatment-naive multiple myeloma (KEYNOTE-185): A randomised, open-label, phase 3 trial. *The Lancet Haematology, 6*, 448–458. https://doi.org/10.1016/S2352-3026(19)30109-7.

Zea, A. H., Rodriguez, P. C., Atkins, M. B., Hernandez, C., Signoretti, S., Zabaleta, J., et al. (2005). Arginase-producing myeloid suppressor cells in renal cell carcinoma patients: A mechanism of tumor evasion. *Cancer Research, 65*(8), 3044–3048. https://doi.org/10.1158/0008-5472.CAN-04-4505.

CHAPTER SEVEN

Multiplexed cytometry for single cell chemical biology

Henry A.M. Schares[a,b,c], Madeline J. Hayes[d,e], Joseph A. Balsamo[f], Hannah L. Thirman[b,c], Brian O. Bachmann[a,b,f], and Jonathan M. Irish[b,c,d,e,g,*]

[a]Department of Chemistry, Vanderbilt University, Nashville, TN, United States
[b]Vanderbilt Institute of Chemical Biology, Nashville, TN, United States
[c]Vanderbilt Chemical and Physical Biology Program, Vanderbilt University, Nashville, TN, United States
[d]Department of Cell and Developmental Biology, Vanderbilt University, Nashville, TN, United States
[e]Department of Pathology, Microbiology and Immunology, Vanderbilt University Medical Center, Nashville, TN, United States
[f]Department of Pharmacology, Vanderbilt University, Nashville, TN, United States
[g]Vanderbilt-Ingram Cancer Center, Vanderbilt University Medical Center, Nashville, TN, United States
[*]Corresponding author: e-mail address: jonathan.irish@vanderbilt.edu

Contents

Methods in Cell Biology, Volume 195
ISSN 0091-679X
https://doi.org/10.1016/bs.mcb.2023.03.007

Abstract

Flow cytometry has great potential for screening in translational research areas due to its deep quantification of cellular features, ability to collect millions of cells in minutes, and consistently expanding suite of validated antibodies that detect cell identity and functions. However, cytometry remains under-utilized in discovery chemical biology due to the differences in expertise between chemistry groups developing chemical libraries and cell biologists developing single cell assays. This chapter is designed to bridge this gap by providing a detailed protocol aimed at both chemistry and biology audiences with the goal of helping train novice researchers. Assay users select from three elements: a small molecule input, a target cell type, and a module of cytometry readouts. For each, we explore basic and advanced examples of inputs, including screening fractionated microbial extracts and pure compounds, and target cells, including primary human blood cells, mouse cells, and cancer cell lines. One such module of cytometry readouts focuses on cell function and measures DNA damage response (γH2AX), growth (phosphorylated S6), DNA content, apoptosis (cleaved Caspase3), cell cycle M phase (phosphorylated Histone H3), and viability (membrane permeabilization). The protocol can also be adapted to measure different functional readouts, such as cell identity or differentiation and contrasting cell injury mechanisms. The protocol is designed to be used in 96-well plate format with fluorescent cell barcoding and the debarcodeR algorithm. Ultimately, the goal is to encourage the next generation of chemical biologists to use functional cell-based cytometry assays in discovery and translational research.

Abbreviations

500x 500-fold more concentrated than the final working concentration of a reagent
c-CAS3 cleaved-caspase3
DMSO dimethyl sulfoxide
FCB fluorescent cell barcoding
IC ice cold, −20 °C
MAM multiplexed activity metabolomics
MAP multiplexed activity profiling
MeOH methanol
p-HH3 phosphorylated Histone H3
PB pacific blue
PBMCs peripheral blood mononuclear cells
PFA para-formaldehyde
PO pacific orange
RT room temperature, 23 °C
UTC uptake control

1. Introduction

Single cell chemical biology approaches monitor multiple aspects of cell biology simultaneously in individual cells and can track bioactivity in mixed populations of cells across time and dose (Krutzik, Crane, Clutter, & Nolan, 2008). Single cell approaches thus reveal differences in responses in subsets of cells that can be obscured in other assays that use an average or sum of the signals (Irish & Doxie, 2014; Krutzik, Irish, Nolan, & Perez, 2004). A major advantage of cellular barcoding is that this approach can be applied in primary tissues or cellular systems containing multiple cell subpopulations that may have contrasting responses to compounds or signaling inputs (Bodenmiller et al., 2012; Brodin et al., 2015; Earl et al., 2018; Irish et al., 2010; Krutzik et al., 2008). However, these protocols are daunting to newer users because they combine techniques from multiple fields and must be executed with precision in the high throughput and multiplexed setting. These challenges have limited the practicality of single cell chemical biology for discovery applications, and we hope to mitigate this issue here by providing a protocol that is generalized for use by a newer student with a basic bench research background.

Fluorescent cell barcoding (FCB) uses differing levels of multiple *N*-hydroxysuccinimidyl ester functionalized amine reactive fluorescent dyes to covalently label cells within a sample (Krutzik & Nolan, 2006). With FCB, each sample has a distinct fluorescent signature allowing samples to be pooled for processing and cytometric single-cell analysis, then demultiplexed to reveal the original populations (Krutzik, Clutter, Trejo, & Nolan, 2011). Barcoding and pooling samples reduces antibody consumption, increases staining uniformity, increases throughput, and decreases cytometer time. The increased throughput and reduced costs enabled by FCB opens the high content multiplexable nature of single cell flow cytometry experiments to an assay format amenable to discovery applications (D'Antonio et al., 2017; Reisman, Barone, Bachmann, & Irish, 2021). Additionally, this method can be utilized in established cancer cell lines (Boyce, Reisman, Bachmann, & Porco, 2021; Manohar et al., 2019) or primary tissues, such as peripheral blood mononuclear cells (PBMC) or cancer patient biopsies (Earl et al., 2018; Giudice, Feng, Kajigaya, Young, & Biancotto, 2017; Irish et al., 2010; Leelatian et al., 2017). In a 48-well barcode screening experiment using a leukemia patient sample and a staining panel with 6 different activity markers and one cell identity marker we can execute 18 separate immunoassays per condition (six activity markers × three differentiated cell types) and 864 individual immunoassays per barcode, all in a single tube (six activity markers × three differentiated cell types × 48 wells). These applications allow for high-throughput multiparametric immunoassay monitoring core cell function(s), cell identity, cell differentiation, cell injury, or chemically induced tumor immunity (CIATI) (Fucikova et al., 2020; Galluzzi, Buque, Kepp, Zitvogel, & Kroemer, 2017; Menard, Martin, Apetoh, Bouyer, & Ghiringhelli, 2008; Michaud et al., 2011; Sriram et al., 2021) in response to chemical challenge.

Despite the benefits gained from FCB, the technique can seem unapproachable to many due to the level of expertise required for manual deconvolution or "debarcoding" of barcoded experiments. To address this barrier of entry to the method, we recently published a computational tool, DebarcodeR, which automatically demultiplexes fluorescent cell barcoded experiments (Reisman et al., 2021; see CytoLab GitHub for DebarcodeR protocol: https://github.com/cytolab/DebarcodeR/tree/master/Protocols). While DebarcodeR addresses a computational barrier to widespread use of single cell chemical biology, here the goal was to address experimental barriers. Thus, the following chapter is intended as an approachable protocol for novices and experts alike who are embarking on a single cell study of compounds, natural products, or other sets of molecules (Fig. 1).

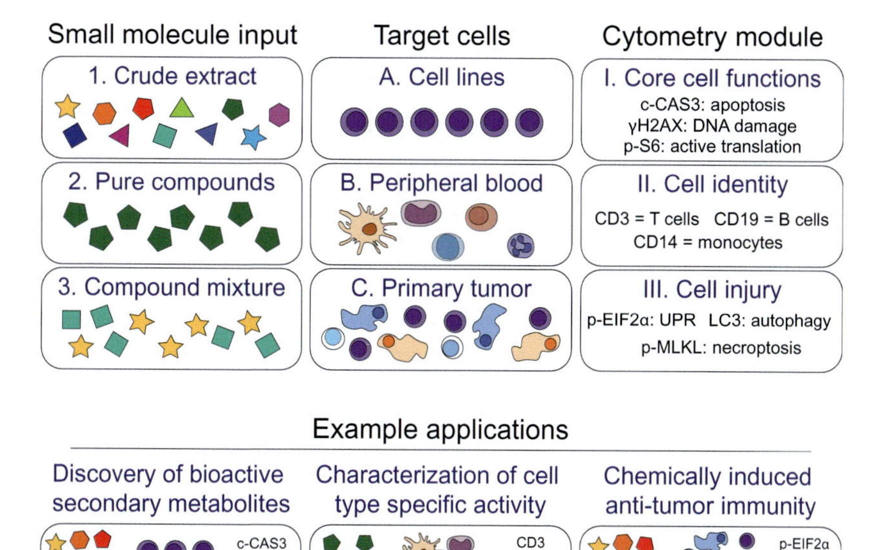

Fig. 1 The single cell chemical biology assay combines small molecule inputs, target cells, and a module of cytometry readouts. Depicted are three core elements of the single cell chemical biology assay that can be changed according to the goals of an experiment. A small molecule input is first selected and can be any chemical entity ranging from crude extracts in a dilution series to chromatographic arrays to individual compounds, as in a traditional high-throughput screening campaign. Next, a set of target cells is selected as the biological model system of interest. Examples can include the MV-4-11 leukemia cell line, human peripheral blood mononuclear cells (PBMCs), or primary human tumor tissue that has been disaggregated into single cells representing blood vessels, immune cells, stroma cells, and malignant cells. Cytometry modules include sets of antibodies, dyes, or other readouts that track cell identity and function. In a typical cytometry panel, 6 to 25 readouts would be measured per cell. Multiplexed staining panels of cell surface and/or intracellular fluorescently conjugated antibodies can be mixed and matched, depending on the application, compatibility of staining conditions, and fluorophore readouts. By picking a set of small molecule inputs, target cells, and cytometry modules, protocol users can dissect mechanisms in cell subtypes or look for cell type specific activity of compound libraries or secondary metabolites.

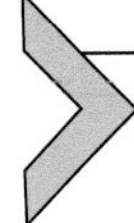

2. Before you begin

2.1 Designing a barcoding experiment

This protocol is intended to be "plug and play," meaning that it should be seen as modular, customizable, and able to support a wide array of applications focused on exploring a large space of variables with a multiplexed flow cytometry assay (Fig. 1). There are three elements that can be adjusted to answer different questions in fields such as immunology and immuno-oncology. The small molecule input for cell challenge is chosen by the investigator and can range from discrete compounds in a drug screening function, to complex cell culture extracts for discovery of novel secondary metabolites or multiplexed dose and time response assays. This protocol has been compatible with all tested established cell lines and heterogeneous cell mixtures, such as peripheral blood mononuclear cells (PBMC) or single cell suspensions of human tumors. Finally, an impactful customizable aspect is the cell function and/or identity staining panel that can be any combination of barcode compatible fluorophores including fluorescent antibodies and functionally diagnostic fluorescent dyes. Barcode-compatible staining panel modules are then applied to cells in order to detect multiple biological signals of interest. Several modules have been established to address a wide range of experimental goals (Fig. 1), and these panels can be modified and mixed depending on application. For example, the module described in the proceeding protocol monitors core cell functions and is used to identify bioactive secondary metabolites. In this application, cancer cell lines are challenged with bacteria-derived metabolomic extracts, fractionated extract, or pure compounds. Upon identification of bioactive molecule, one can modify or switch both modules and cell input type. For example, a secondary assay might investigate cell injury pathways that are implicated in eliciting the initially observed perturbation to core cell functions. Indeed, the compatibility of barcode dyes and fluorescent readouts enables applications in heterogenous cell populations, such as primary human tissues in single cell suspension. Barcode dyes and modules can be coupled with additional fluorescence compatible cell identity markers to define activity profiles of discrete immune cell populations and establish nascent evidence for compound specific cell subtype selectivity.

2.2 Assay setup

The protocol below describes the production of a single 48-well barcode experiment. However, the assay is optimized for execution in 96-well plates containing two separate 48-well barcoded experiments, each with their own set of controls. Additionally, depending on the experience level of the individual carrying out the assay, multiple 96-well plates can be processed together. On a typical day of executing this protocol, an individual might run 4 total 96-well plates, collecting 384 wells and millions of cells in just 8 tubes of 48 barcoded wells. While a single session of the protocol with 4 plates can be executed in time for sensitive processing steps and to prevent time-dependent variation in protocol execution, we suggest starting with fewer plates and a single cell type while learning.

3. Materials

3.1 Stocks of buffers and dyes

1. PBS/BSA—1% (w/v) BSA in PBS.
   - **(a)** Let 5 g BSA dissolve in 500 mL PBS and sterile filter.
   - **(b)** Store at 4 °C and warm to room temperature (RT) before using.
2. 500× dimethyl sulfoxide (DMSO) stock of AlexaFluor 700 (20 μg/mL).
   - **(a)** Add 1 mg Ax700 to 50 mL DMSO and pipette to mix.
   - **(b)** Aliquot into desired volume and store at −80 °C.
   - **(c)** Let thaw completely at RT before using.
3. 500× DMSO stock of AlexaFluor 750 (500 μg/mL).
   - **(a)** Add 1 mg Ax750 to 2 mL DMSO and pipette to mix.
   - **(b)** Aliquot into desired volume and store at −80 °C.
   - **(c)** Let thaw completely at RT before using.
4. 500× DMSO stock of PO (500 μg/mL).
   - **(a)** Add 1 mg PO to 2 mL DMSO and pipette to mix.
   - **(b)** Aliquot into desired volume and store at −80 °C.
   - **(c)** Let thaw completely at RT before using.
5. 500× DMSO stock of PB (500 μg/mL) in DMSO.
   - **(a)** Add 5 mg PB to 10 mL DMSO and pipette to mix.
   - **(b)** Aliquot into desired volume and store at −80 °C.
   - **(c)** Let thaw completely at RT before using.

6. Cell Media: IMDM containing L-glutamine and 25 mM HEPES, 10% (v/v) fetal bovine serum, and antibiotics.
   (a) Combine 450 mL IMDM, 50 mL FBS, and 5 mL Penicillin-Streptomycin and sterile filter.
   (b) Keep in 4 °C and warm to 37 °C before using.

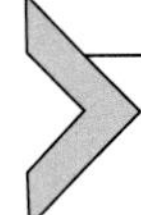

4. Key resources table

Reagent or resource	Source	Identifier
Antibodies		
PE Rabbit Anti-Active Caspase-3 (c-CAS3), clone: C92-605	Fisher	BD550821
H2AX (p2139) Mouse, PerCP-Cy5.5, Clone: N1-4312	Fisher	BD 564718
Phospho-S6 Ribosomal Protein (Ser235/236) (D57.2.2E) XP→Rabbit mAb (Alexa Fluor→647 Conjugate	Cell Signaling Technology	4851
PE/Cyanine7 anti-Histone H3-Phosphorylated Ser28 (p-HH3)	BioLegend	641011
Chemicals, peptides, and recombinant proteins		
Methanol (MeOH)	Fisher	A412-4
Dimethyl sulfoxide (DMSO)	Fisher	BP231-1
16% Paraformaldehyde	Fisher	50980487
Dyes		
Pacific orange (PO) succinimidyl ester, triethylammonium salt	ThermoFisher	P30253
Pacific blue (PB) succinimidyl ester	ThermoFisher	P10163
Alexa fluor 750 NHS ester (succinimidyl ester)	ThermoFisher	A20111
Alexa fluor 700 NHS ester (succinimidyl ester)	ThermoFisher	A20010
YO-PRO-1 Iodide (491/509)	ThermoFisher	Y3503
Experimental models: Cell lines		
MV-4-11	ATCC	CRL-9591
Small molecules		
Staurosporine	LKT Labs	S7600

—cont'd

Reagent or resource	Source	Identifier
Etoposide	Cayman Chemicals	33419-42-0
Aphidicolin	Fisher	AC611970010
Rapamycin	LC Labs	NC9163747
Nocodazole	Acros	358240100

5. Materials and equipment

- 96-Well Tissue Culture Plate (Fisher, 229196)
- 96-Well Non-Treated V-Bottom Polypropylene Microplate (Corning, 353263)
- Aluminum 96-Well Plate Seal (VWR, 60941-126)
- 2 mL Microcentrifuge Tubes (Fisher, 69720)
- 15 mL Conical Centrifuge Tube (Fisher, 14-959-49B)
- 50 mL Conical Centrifuge Tube (Fisher, 14-432-22)
- Round-Bottom Polystyrene Test Tube, FACS tube (Fisher, 14-959-5)
- Phosphate Buffered Saline, PBS (Fisher, MT21040CV)
- Bovine Serum Albumin, BSA (Millipore Sigma, A4503)
- RNASE A (Qiagen, 19101)
- AbC Total Antibody Compensation Bead Kit (Invitrogen, A10497)
- IMDM containing L-glutamine and 25 mM HEPES (Gibco, 12440053)
- Fetal Bovine Serum (Corning, MT35010CV)
- Penicillin-Streptomycin (Fisher, 15140122)
- Centrifuge compatible with spinning plates
- 37 °C, 5% CO_2 incubator

Optional:

- Multichannel pipettor, e.g., Pipet-Lite Multi Pipette L12-200XLS+ (Rainin, 17013810)

6. Prepare plates

6.1 Prepare barcoding plates

Timing: 1–2 h

1. Pacific blue (PB) eight-level barcode preparation.
   - **(a)** Label one 15 mL falcon tube (or tube with similar composition with >2 mL capacity) PB8 and label seven 2 mL microcentrifuge tubes PB7-PB1.

  (b) Dispense 2540 μL DMSO to PB8 and 710 μL DMSO to PB7-PB1.
  (c) Add 52 μL of PB (500 μg/mL stock concentration) to PB8 triturating (pipetting up and down) at least five times to mix, forming PB level 8 at 10 μg/mL.
  (d) Transfer 1 mL from PB8 to PB7, triturating at least five times to mix, forming PB level 7.
  (e) Transfer 1 mL from PB7 to PB6, triturating at least five times to mix, forming PB level 6.
  (f) Repeat transfer of 1 mL between levels to the formation of PB level 1.
2. Pacific Orange (PO) 6-level barcode preparation.
  (a) Label one 15 mL falcon tube (or tube with similar composition with >2 mL capacity) PO6 and label five 2 mL microcentrifuge tubes PO5–PO1.
  (b) Dispense 3011 μL of DMSO to PO6 and 710 μL of DMSO to PO5–PO1.
  (c) Add 262 μL of PO (500 μg/mL stock concentration) to PO6 triturating at least five times to mix, forming PO level 6 at 40 μg/mL.
  (d) Transfer 1 mL from PO6 to PO5, triturating at least five times to mix, forming PO level 5.
  (e) Transfer 1 mL from PO5 to PO4, triturating at least five times to mix, forming PO level 4.
  (f) Repeat transfer of 1 mL between levels to the formation of PO level 1.
3. AlexaFluor 750 internal control preparation.
  (a) In a 15 mL falcon tube (or tube with similar composition with >8 mL capacity), dispense 7707 μL of DMSO.
  (b) Add 78 μL of AlexaFluor 750 Succinimidyl Ester (500 μg/mL stock concentration), triturating at least five times to mix for a final concentration of 5 μg/mL.
4. Preparation of 48-well fluorescent cell barcode primary plate.
  (a) Transfer 90 μL of 5 μg/mL AlexaFluor 750 to all 48 wells onto the first half of a 96-well v-bottom polypropylene plate (Rows A–H, Columns 1–6).
  (b) Transfer 90 μL of PB8 to row A, columns 1–6.
  (c) Transfer 90 μL of PB7 to row B, columns 1–6.
  (d) Transfer 90 μL of PB6 to row C, columns 1–6.
  (e) Repeat this pattern 5 more times ending with transferring 90 μL of PB1 to row H, columns 1–6.
  (f) Transfer 90 μL of PO6 to column 1, row A–H.

(g) Transfer 90 μL of PO5 to column 2, row A–H.
(h) Transfer 90 μL of PO4 to column 3, row A–H.
(i) Repeat this pattern three more times ending with transferring 90 μL of PO1 to column 6, row A–H.
(j) All 48 wells (Rows A–H, Columns 1–6) should have 270 μL once completed (90 μL Ax750, 90 μL PB, 90 μL PO).
(k) Spin the plate down in a centrifuge at 800 × *g* for 5 min and seal with adhesive aluminum sealing film for storage at −80 °C or proceed directly to secondary plate preparation.
(l) Leftover dye from the highest level should be split into 6 μL single dye aliquots in 2 mL microcentrifuge tubes. These will be used for **COMP_TUBE_Ax750**, **COMP_TUBE_PO**, **COMP_TUBE_PB** tubes later in the protocol. These microcentrifuge tubes can be kept at −80 °C for up to 3 months.
(m) In addition, for each 48-well barcoding set prepared, one **UTC_CELLS** tube must be prepared with leftover dye by adding 5 μL each of the highest levels of PO (PO6), PB (PB8), and Ax750 into one 2 mL microcentrifuge tube. **UTC_CELLS** dye tubes can be kept at −80 °C alongside plates.

5. Secondary plate preparation.
(a) Transfer 15 μL from each primary plate well to corresponding secondary plate wells as well as the second half of the 96-well plate so that you have two barcodes per secondary plate (A1 to A1 and A7, A2 to A2 and A8, etc.).
(b) Seal secondary plate with adhesive aluminum sealing film and store at -80 °C for up to 3 months.
(c) The secondary plate is referred to as **BARCODE_PLATE** from here on.

Note: This preparation generates 30 sets of 48 barcodes in an 8 by 6 configuration. Once transferred to 96-well plates, this will fill 15 plates with two sets of 48-well barcodes per plate. The barcodes will use three cytometer channels: two channels for PB and PO and the Ax750 channel for the uptake control (UTC).

6.2 Prepare a compound plate

Timing: Timing varies by preferred preparation method 0.5–1 h

1. For the outlined staining panel, we include three vehicle controls and five positive controls delivered in 1 μL of DMSO that we place in the far-right column of the barcode (in a 96-well plate containing two 48-well barcodes: columns 6 and 12).

2. A6: 1 μL DMSO (Final Concentration: 0.5% DMSO).
3. B6: 1 μL DMSO (Final Concentration: 0.5% DMSO).
4. C6: 1 μL DMSO (Final Concentration: 0.5% DMSO).
5. D6: 1 μL 200 μM Staurosporine (Final Concentration: 1 μM)—c-CAS3 and γH2AX positive control.
6. E6: 1 μL 2000 μM Etoposide (Final Concentration: 10 μM)—γH2AX and c-CAS3 positive control.
7. F6: 1 μL 800 μM Aphidicolin (Final Concentration: 4 μM)—G_1 cell cycle arrest positive control.
8. G6: 1 μL 2 μM Rapamycin (Final Concentration: 0.01 μM)—p-S6 suppression positive control.
9. H6: 1 μL 800 μM Nocodazole (Final Concentration 4 μM)—G_2 cell cycle arrest and p-HH3 positive control.
10. For the remaining 40 available wells, conditions vary and should be plated alongside controls listed above in one 96-well tissue culture plate. Once all 48-wells of a plate (rows A–H, columns 1–6) have a condition set this plate is referred to as the **COMPOUND_PLATE**.

Note: The contents of the **COMPOUND_PLATE** can be varied including the controls, but wells should be reserved for multiple vehicle controls and a positive control condition for each antibody stain when designing a **COMPOUND_PLATE**.

Critical: Make sure DMSO concentration is consistent across all conditions either by dosing compounds in DMSO or normalizing DMSO concentration when adding cells.

Pause Point: The protocol can be paused indefinitely after preparing the barcoding and compound plates. Plates should be stored at -80 °C until ready to use.

7. Prepare cells

7.1 Plate cells and treat with compounds

Timing: 20–30 min

1. Acquire 10–20 million MV-4-11 cells from suspension culture ranging from 0.5 to 1.0 million cells/mL.
2. Pellet the cells by centrifuging at 200 × *g* for 5 min.
3. Decant supernatant and resuspend in media into single cell culture to a cell density of around 500,000 cells/mL.

4. Set aside half of the cell suspension in a separate flask to be used as **UNSTAINED_CELLS** for preparation of compensation controls.
5. Dispense 200 μL of cell suspension (100,000 cell/well) to each of the 48 wells of the **COMPOUND_PLATE**, pipetting to mix.
   (a) Now that cells have been added the **COMPOUND_PLATE** will be referred to as the **TREATED_PLATE**.
6. Incubate **TREATED_PLATE** and **UNSTAINED_CELLS** at 37 °C, 5% CO_2 for 16 h or desired treatment time.

Note: The number of cells required to carry out an experiment will change if performing multiple barcodes simultaneously. Be sure to calculate how many cells you will need prior to starting the protocol.

7.2 Stain cells for viability, then fix and permeabilize

Timing: 1–2 h

1. Prepare 10× Ax700 solution (0.4 μg/mL) by adding 60 μL of 500× Ax700 stock (20 μg/mL) into 3 mL PBS.
2. Add 21 μL of the 10× Ax700 solution to each well of the **TREATED_PLATE**.
   (a) Also freeze or set aside 21 μL 10× Ax700 in a 2 mL microcentrifuge tube alongside the **COMP_TUBE_[stain]** tubes from **BARCODE_PLATE** preparation for later use.
3. Incubate **TREATED_PLATE** for 5 min at 37 °C.
   (a) During incubation, prepare **FIX_PLATE** by adding 25 μL 16% paraformaldehyde (PFA) to each of the 48 wells of a new 96 well polypropylene v-bottom plate to match the layout of the **TREATED_PLATE**.
4. Transfer contents of the **TREATED_PLATE** to the **FIX_PLATE** well to well (A1 to A1, B2 to B2, etc.), triturating three times to mix while transferring.
   (a) Discard the **TREATED_PLATE** once transfer is complete.
5. Incubate the **FIX_PLATE** for 10 min at RT in the dark.

Note: Beware of over fixing. Do not let any single plate incubate for too long with PFA.

6. Once the 10-min incubation has concluded, centrifuge the **FIX_PLATE** at 800 × *g* for 5 min, flick the supernatant to decant, and vortex the plate in void volume to break up the pellet.

Note: When vortexing, ensure cell pellets are broken up and resuspended in the void volume to prevent aggregation and formation of doublets. Fixed cells should be able to withstand vigorous vortexing.

7. To permeabilize cells, resuspend each well in 200 μL of 100% ice cold (IC) methanol (MeOH) and triturate three to five times to completely resuspend cells.

Critical: It is vital that the MeOH and the suspension of cells in MeOH maintain IC temperature; this step and the next should be completed without delay.

8. Store the **FIX_PLATE** at −20 °C for at least 30 min or longer.

Critical: To maintain IC temperature, immediately transfer cells in MeOH to a −20 °C freezer until continuing the protocol.

Pause Point: The protocol can be paused indefinitely after the cells are in MeOH. Fixed and permeabilized cells are stable in MeOH if sealed and kept dry and can be stored at −20 °C for 1–2 weeks or −80 °C for up to a month.

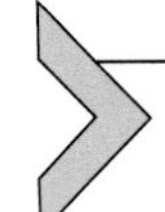

8. Barcode cells and antibody stain

8.1 Barcode cells with dyes

Timing: 1–2 h

1. Take a **BARCODE_PLATE** out of the freezer and thaw at RT for at least 30 min before beginning.
2. Also remove a **UTC_CELLS dye** tube from the freezer to thaw.
3. Centrifuge the **FIX_PLATE** and thawed **BARCODE_PLATE** at 800 × *g* for 5 min.
4. Remove the adhesive seal from the thawed **BARCODE_PLATE** and set aside for now without decanting.
5. Flick the **FIX_PLATE** to decant supernatant and vortex in void volume.
6. Resuspend each well of the **FIX_PLATE** in 205 μL PBS.
7. Triturate the contents of the **FIX_PLATE** three times to resuspend and transfer 185 μL from the **FIX_PLATE** to the **BARCODE_PLATE** well to well (A1 to A1, B1 to B1, etc.) triturating to mix.

Note: Important to triturate at least three times in the **FIX_PLATE** to ensure plate contents are a single cell suspension before transferring plates or cells may not be carried over.

8. Incubate the **BARCODE_PLATE** for 30 min at RT in the dark.
   - **(a)** During the incubation, pool the cells remaining in the **FIX_PLATE** and transfer into one FACS tube labeled **Ax700_CELLS**.
   - **(b)** Add 1 mL PBS/BSA to **Ax700_CELLS**.
   - **(c)** Centrifuge at 800 × *g* for 5 min, flick supernatant to decant, and vortex in void volume.

   (d) Resuspend **Ax700_CELLS** in 300 μL PBS and vortex.
   (e) Transfer 185 μL from **Ax700_CELLS** into thawed **UTC_CELLS** dye tube containing pre-made dye mixture, vortex tube, transfer the entire volume to a FACS tube labeled **UTC_CELLS**, then incubate in the dark for 30 min.
   (f) **Ax700_CELLS** is done; set aside in the dark.
9. After the **BARCODE_PLATE** has incubated for 30 min, quench by adding 70 μL PBS/BSA to each well.
10. Centrifuge **BARCODE_PLATE** at 800 × *g* for 5 min, flick supernatant to decant, vortex in void volume.
11. After **UTC_CELLS** incubates for 30 min, quench by adding 500 μL PBS/BSA to the FACS tube.
12. Centrifuge **UTC_CELLS** at 800 × *g* for 5 min, flick supernatant to decant, vortex in void volume.
13. Resuspend in 200 μL PBS/BSA. **UTC_CELLS** is done; set aside in the dark.

Note: Be wary of the timing of the two different 30 min incubations—it should overlap that **UTC_CELLS** finishes while **BARCODE_PLATE** is spinning in the centrifuge.

8.2 Stain cells with antibodies

Timing: 1–2 h

1. Resuspend *Row A* of **BARCODE_PLATE in** 200 μL PBS/BSA.
2. Use the 200 μL PBS/BSA in each well of Row A to collect and pool the contents of all wells by triturating in all wells of the column to pool contents on the plate in Row H.
3. Transfer all wells in Row H to a FACS tube labeled **FCB_CELLS**.
4. Centrifuge **FCB_CELLS** at 800 × *g* for 5 min, flick supernatant to decant, vortex in void volume.
5. Resuspend **FCB_CELLS** in 100 μL PBS/BSA.
6. Make **STAINING_MIX** (per 48-well barcoding experiment): Add 10 μL c-CAS3-PE, 2.5 μL γH2AX-PerCP-Cy5.5, 0.25 μL p-S6-Ax647, and 0.1 μL p-HH3-PE-Cy7 to a microcentrifuge tube, pipetting to mix.
7. Transfer 87 μL from **FCB_CELLS** to a FACS tube labeled **STAINED_CELLS**.
   (a) Add an additional 87 μL of PBS/BSA to **FCB_CELLS** and vortex.
   (b) **FCB_CELLS** is done; set aside in the dark.

8. Add 13 μL of **STAINING_MIX** to **STAINED_CELLS**, vortex, and incubate in the dark at RT for 30 min or 4 °C overnight.
9. After staining is complete, wash **STAINED_CELLS** twice by adding 1 mL PBS/BSA, centrifuging at 800 × *g* for 5 min, flick supernatant, vortex in void volume, and repeat.
   (a) During centrifugations, add 1 μL of 1 mM YO-PRO, 40 μL of RNase A, and 959 μL of PBS/BSA to a microcentrifuge tube to form a working concentration of 1 μM YO-PRO (50×).
10. After washes are complete, resuspend **STAINED_CELLS** in 490 μL PBS/BSA.
11. Add 10 μL of working YO-PRO to **STAINED_CELLS** and vortex. **STAINED_CELLS** is done; set aside in the dark.
   (a) Also set aside 10 μL working YO-PRO in a 2 mL microcentrifuge tube alongside the **compensation dyes in microcentrifuge** tubes from **BARCODE_PLATE** preparation for later use.

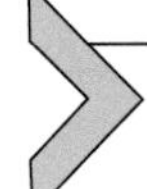

9. Prepare cytometry controls

9.1 Create compensation controls

Timing: 2–3 h—This can be performed as described or in tandem with the treated plate

1. Take microcentrifuge tubes containing aliquots of PO, PB, and Ax750 prepared during **BARCODE_PLATE** preparation out of the freezer and thaw at RT for at least 30 min before beginning.
2. For compensation controls label 10 FACS tubes:
   (a) **COMP_TUBE_PO**
   (b) **COMP_TUBE_PB**
   (c) **COMP_TUBE_Ax750**
   (d) **COMP_TUBE_Ax700**
   (e) **COMP_TUBE_PE**
   (f) **COMP_TUBE_PerCPCy5.5**
   (g) **COMP_TUBE_Ax647**
   (h) **COMP_TUBE_PECy7**
   (i) **COMP_TUBE_UNSTAINED_CELLS**
   (j) **COMP_TUBE_UNSTAINED_BEADS**
3. Transfer **UNSTAINED_CELLS** from flask to a 50 mL falcon tube.
4. Add 10% v/v of 16% PFA to **UNSTAINED_CELLS** (e.g. 500 μL PFA into 5 mL cells).
5. Incubate for 10 min at RT in the dark.

6. While the **UNSTAINED_CELLS** fix, add 5 μL PO from previously prepared aliquot to **COMP_TUBE_PO**, 5 μL PB from previously prepared aliquot to **COMP_TUBE_PB**, 5 μL AlexaFluor 750 from previously prepared aliquot to **COMP_TUBE_Ax750**, 21 μL 10× Ax700 to **COMP_TUBE_Ax700**, and 5 μL AlexaFluor 750 to **COMP_TUBE_Ax750**.
7. Centrifuge **UNSTAINED_CELLS** at 800 × *g* for 5 min, flick supernatant to decant, and vortex in void volume.
8. To permeabilize cells, resuspend in the same volume as step 3 of 100% ice-cold MeOH (e.g., 500 μL).
9. Store **UNSTAINED_CELLS** at −20 °C for at least 30 min.

Critical: To maintain antigen structures, it is vital the MeOH remains cold (−20 °C) when dispensing and transfer of cells to −20 °C freezer take place immediately to maintain temperature.

Pause Point: The protocol can be paused indefinitely once the cells are in MeOH. Fixed and permeabilized cells are stable in MeOH if sealed and kept dry and can be stored at −20 °C for 1–2 weeks or −80 °C for months or more.

10. Centrifuge **UNSTAINED_CELLS** at 800 × *g* for 5 min, flick supernatant to decant, vortex in void volume.
11. Resuspend in 2 mL of PBS.
12. Transfer 195 μL of **UNSTAINED_CELLS** to **COMP_TUBE_PO**, **COMP_TUBE_PB**, **COMP_TUBE_Ax750** and 200 μL of **UNSTAINED_CELLS** to **COMP_TUBE_Ax700**.
13. Vortex all tubes once cells have been added.
14. Incubate all four tubes in the dark for 30 min.
15. Quench all four tubes after 30 min by adding 500 μL PBS/BSA to each tube.
16. Centrifuge at 800 × *g* for 5 min, flick supernatant to decant, vortex in void volume.
17. Resuspend each tube in 200 μL PBS/BSA. **COMP_TUBE_PO**, **COMP_TUBE_PB**, **COMP_TUBE_Ax700**, and **COMP_TUBE_Ax750** are done. Set them aside in the dark.
18. Add 10 μL working YO-PRO, 100 μL **UNSTAINED_CELLS** and 390 μL PBS/BSA to **COMP_TUBE_YO-PRO** and vortex. **COMP_TUBE_YO-PRO** is done. Set it aside in the dark.
19. Transfer remaining **UNSTAINED_CELLS** to **COMP_TUBE_UNSTAINED_CELLS**. **COMP_TUBE_UNSTAINED_CELLS** is done. Set it aside in the dark.

20. Prepare **COMP_TUBE_PE**, **COMP_TUBE_PerCPCy5.5**, **COMP_TUBE_Ax647**, and **COMP_TUBE_PECy7** compensation controls by adding 200 μL PBS/BSA, 1 μL of each corresponding antibody, 10 μL positive capture beads, and 10 μL negative capture beads and vortex. **COMP_TUBE_PE**, **COMP_TUBE_PerCPCy5.5**, **COMP_TUBE_Ax647**, and **COMP_TUBE_PECy7** are done. Set them aside in the dark.
21. Prepare unstained bead tube **COMP_TUBE_UNSTAINED_BEADS** by adding 200 μL PBS/BSA, 1 drop positive capture beads, and 1 drop negative capture beads and vortex. **COMP_TUBE_UNSTAINED_BEADS** is done. Set it aside in the dark.
22. Once all tubes are complete proceed to run on flow cytometer to collect data.

Note: Flow cytometer set up and operation will be specific to a given institution. The unstained tubes and compensation tubes should be run first as part of flow cytometer set up, followed by the experimental tubes: **Ax700_CELLS**, **UTC_CELLS**, **FCB_CELLS**, and **STAINED_CELLS**. If multiple barcoding experiments are performed simultaneously, only one set of compensation tubes are needed, but each barcoding experiment will require separate **Ax700_CELLS**, **UTC_CELLS**, **FCB_CELLS**, and **STAINED_CELLS** tubes. The **STAINED_CELLS** tube is the primary experimental data. and the **Ax700_CELLS**, **UTC_CELLS**, and **FCB_CELLS** are used in debarcoding and troubleshooting.

23. Optimize compensations. For a guide to compensations see cited resource (Roederer, 2002).
24. Apply gating scheme (Fig. 2A).
25. Debarcode data using DebarcodeR per published instructions (Reisman et al., 2021) and the accompanying protocol on GitHub: https://github.com/cytolab/DebarcodeR/tree/master/Protocols.

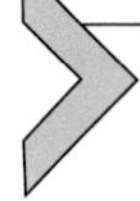

10. Expected outcomes

10.1 Fluorescent cell barcode quality

As a quality control check prior to debarcoding, it is advised to check the quality of the barcode by viewing the data from **STAINED_CELLS** with PB as the x-axis and PO as the y-axis. Examples of barcoded homogenous cancer cell lines, healthy cells, and heterogeneous human blood cells run in parallel and treated with identical compound plates, prior to debarcoding are

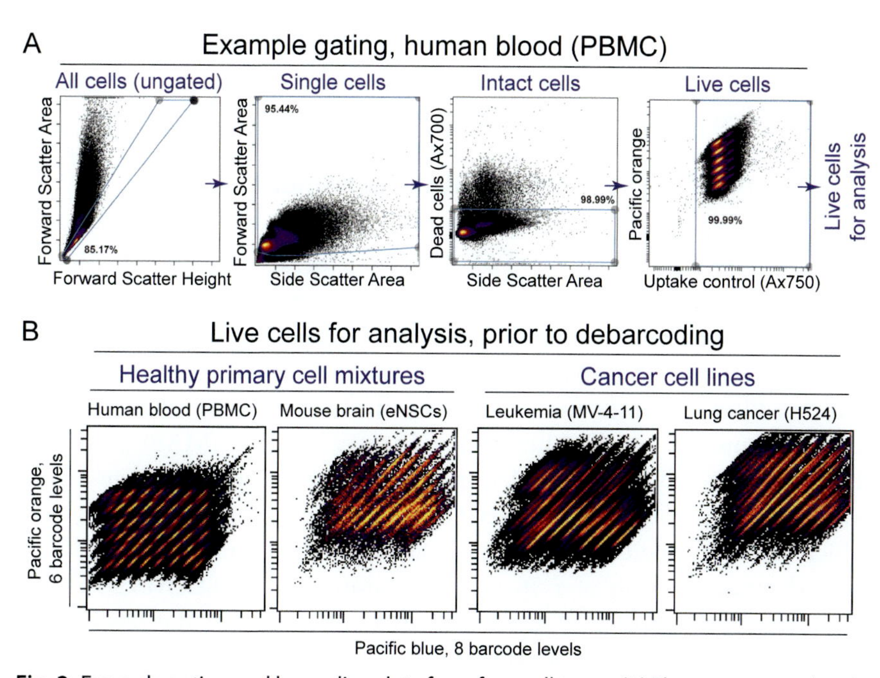

Fig. 2 Example gating and barcoding data from four cell types. (A) Flow cytometry density dot plots show an example gating scheme using data from healthy human PBMCs. First, forward scatter area and height are used to identify single cells. Next, forward scatter and side scatter are used to gate for intact cells. After that, dead cells that take up Ax700 dye are excluded. Finally, within the $Ax700^{-}$ live cells, a gate is drawn to analyze only those cells that stain with Ax750. This gate will be used to define the UTC, a key control used in the debarcodeR process. (B) Once gated as in (A), the live cells for analysis can be visualized by plotting pacific orange and pacific blue. The density dot plots here show a clear 6x8 grid for healthy human PBMCs, which are generally a uniform population of small cells. Next, mouse embryonic neural stem cells are shown. As compound impacted these cells, some wells in the upper left are missing many cells. Similarly, human MV-4-11 leukemia cells also displayed responses in wells here. Finally, small cell lung cancer cell line H524 displayed fewer responses and so most wells were present. The large spread in sizes of H524s means that they pick up more of the barcoding dyes and can appear as diagonal streaks in this view. These correlations between cell size, dye uptake, and barcodes are used by debarcodeR to assign cells to wells and debarcode the data.

shown in Fig. 2 to help illustrate how variable cell types and chemical challenge can impact barcoding. Generally, barcodes in this view should resemble a rectangle made up of a set of eight by six ellipses that may appear as distinguishable populations, as in the case of the PBMC example barcode (Fig. 2), or as diagonal lines, as with H524 cells that vary greatly in size and thus appear less separated in 2D despite being well-separated when considering other parameters, including the uptake control. The more active the small molecule input, the more distorted the barcode may seem as exemplified by MV-4-11s in Fig. 2.

10.2 Example assay output: MAM experiment

A frequently used application of FCB by our group is in the discovery of bioactive microbial secondary metabolites, the process of which exemplifies the extent of the customizability of the assay platform. In the initial stages of the discovery process, prioritization of active extracts is achieved by assaying leukemia cell lines (e.g., MV-4-11) response to microbial culture extracts plated in dilution series and evaluated for perturbation of core cell functional markers, a process termed Multiplexed Activity Profiling (MAP). Extracts deemed sufficiently active in MAP are further investigated using an assay configuration termed Multiplexed Activity Metabolomics (MAM), in which the small molecule input is an active extract chromatographically arrayed across 36 wells of a 96-well plate using UV-HPLC/MS so each chromatographic minute of UVA/MS data correspond to the secondary metabolites that elute into a single well (Earl et al., 2018). A replication of the initial MAP extract series dilution is included along with functional marker compound positive controls and vehicle replicate negative controls. The experiment is carried out as described and the result is as shown (Fig. 3). Individuals may wish to visualize the completed experiment in different ways. Our example graphical visualization of the data for assessment of data quality and analysis of cell perturbation is shown (Fig. 3). In this example, each dot represents a single cell assigned to its original well position by debarcodeR. Black dots originate from **STAINED_CELLS** and indicate marker specific fluorescent signal intensity (Fig. 3A, rows). Red dots originate from **FCB_CELLS** and indicate marker specific background fluorescence of single cells that have not been stained other than with barcoding dyes. These data are useful to observe fluorescent interference from added small molecules and account for cell type specific fluorescent background. Each column constitutes an 'activity profile' of the small molecule input contained in a single well, where each row is a different single cell readout of the same cell population (plate well). Marker shifts should only occur in the black dots and are compared to vehicle wells and positive controls to determine hits.

These data led to the isolation and identification of a bioactive natural product. A hypogean microbial metabolomic extract eliciting strong DNA damage activity inferred via increased phosphorylation of γH2AX,

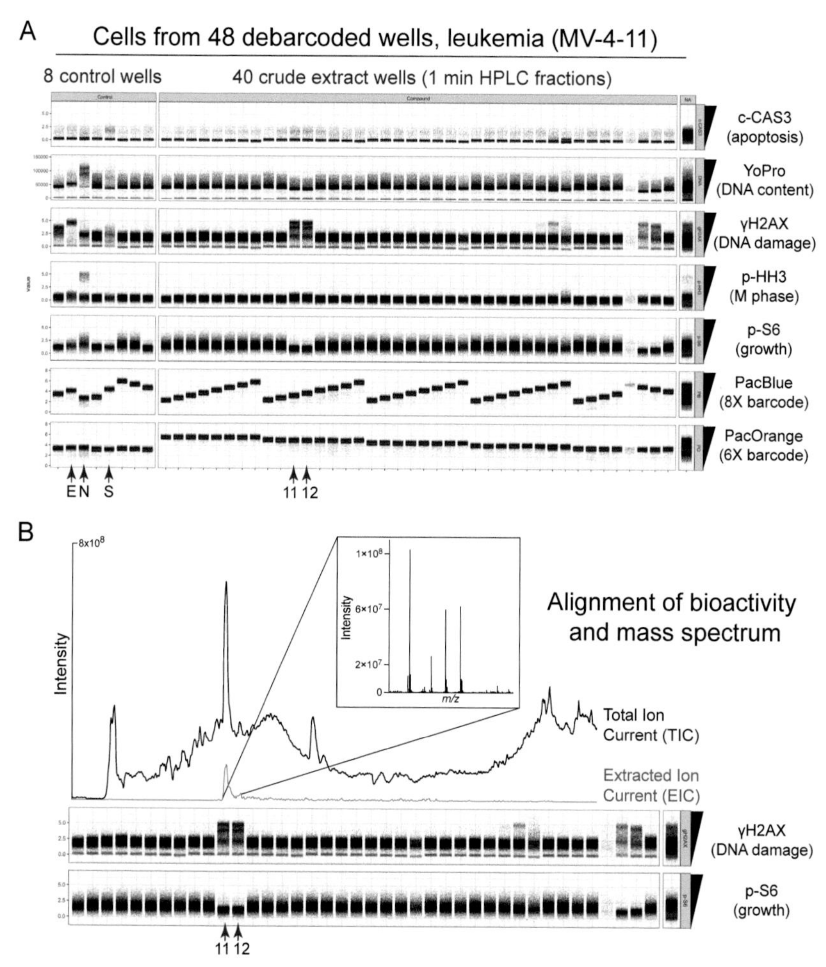

Fig. 3 Example output of the protocol after debarcoding. (A) Output from DebarcodeR is shown. Shown first are 8 control wells containing compounds like etoposide (E), nocodazole (N), or staurosporine (S), positive controls for γH2AX, p-HH3, and c-CAS3, respectively. Next, 40 wells from a chromatographically arrayed hypogean actinomycete crude extract are shown. In this example, the extract in wells 11 and 12 increased γH2AX, a sign of a DNA damage response, and decreased p-S6, a sign of impaired translation and growth suppression. While all rows are measured simultaneously in each cell, the view here splits them out as one per row to help distinguish activity profiles. Each row shows the readout for each core cell function marker and barcoding dye of the cells within a given well (c-CAS3—apoptosis; DNA = cell cycle status; γH2AX—DNA damage; p-HH3—mitosis; p-S6—protein synthesis, growth, and mTOR signaling; PB—Pacific Blue; PO—Pacific Orange). Each column constitutes the 'activity profile' of the contents of a well. Single cell data is transformed using the arcsinh transformation, as is standard for flow cytometry, and data are plotted so that increasing signal is on the y-axis. Black dots represent fluorescence levels of single cells from **STAINED_CELLS** that are treated

(Continued)

suppression of active translation and growth inferred via decreased phosphorylation of p-S6, and cell death inferred by the loss of cells (these cells are AlexaFluor 700 positive and are gated out in the quality control processing, Fig. 2A) in MAP. Initial MAP data are not shown, but replication of the original MAP screening conditions is included for the MAM compound plate as a positive control. To identify the secondary metabolite eliciting the observed activity, the active extract was further investigated via MAM (Fig. 3A). MAM results revealed the same γH2AX and p-S6 marker shifts observed in the initial MAP screen in wells 11 and 12 of the MAM plate (Fig. 3A). Upon examination of UV and mass spectra corresponding to wells 11 and 12, the bioactive secondary metabolite was able to be identified and isolated using Sephadex LH20 resin column and reverse-phase preparative HPLC (Fig. 3B). The secondary metabolite is currently being investigated further.

11. Advantages

Contemporary chemical biology studies aiming to understand and manipulate complex human diseases require assaying multiple cell status markers, at single cell resolution, in heterogenous cell mixtures under multiple conditions. Flow cytometry is particularly powerful in this regard in that it is a high dimensional single cell assessment technique that may be used to simultaneously measure perturbations in multiple surface cell identity and intracellular functional markers on millions of cells in a biological sample. However, limitations of traditional flow cytometry include requiring expensive antibody reagents and significant cytometer and data analysis time per-sample. FCB and DebarcodeR directly addresses these limitations and

Fig. 3—Cont'd with compound, barcoded, and stained with the module staining mix. Red dots represent fluorescence levels of single cells from **FCB_CELLS** that are treated with compound and barcode. Red dots are useful as a non-stained baseline to ensure observed marker shifts are in fact due to metabolite activity and not interfering metabolite autofluorescence. POL, Pacific Orange Level where POL6 is the highest PO Level and POL1 is the lowest. PBL, Pacific Blue Level where PB8 is the highest PB Level and PB1 is the lowest. (B) Positive mode total ion current (TIC) of the actinomycete metabolome and selected extracted ion current (EIC) of the bioactive secondary metabolite where each chromatographic minute corresponds to the contents of each experimental well. Extracted mass spectrum shows the observed metabolites eluted into the active well.

direct benefits are apparent in substantially reduced reagent consumption, analysis time, and intra-well error. Together, the benefits of FCB and DebarcodeR enable highly controlled, modular, high-throughput, high-content single cell chemical biological studies.

11.1 High-throughput sample multiplexing flow cytometry

Sample multiplexing enabled by FCB permits the simultaneous analysis of hundreds of samples using two or more barcoding dyes. Herein we describe a protocol using two dyes to allow pooling of 48 microtiter well samples and simultaneous assay of multiple timepoints, doses, and compounds against several functional readouts. The result is a substantial improvement compared to high-throughput targeted assays. A single flow cytometric run generates 288 individual well based immunoassays (six functional assays per well) in cancer cell lines, and over 864 individual immunoassays when challenging a heterogeneous PBMC sample containing three distinguishable cell types. Given that each sample contains thousands of cells per well, the resulting response profile captures a range of cell responses for a given condition, more accurately reflecting cellular response phenotypes than aggregated well measurements. A typical set of experiments for a single user in our lab involves measuring four 96-well plates allowing for eight sets of 48-well barcoding or 384 conditions in one experiment. This comes out to 2,304 individual immunoassays using cell lines, and almost 7,000 individual immunoassays using a heterogeneous cell sample with three distinguishable cell types. Notably, if additional barcoding dyes are employed, sample multiplexing can in principle be increased from 48 to thousands of samples and tens of thousands of assays per flow cytometric run.

11.2 Cost analysis of FCB

A significant advantage of pooling barcoded samples prior to antibody staining is a substantial decrease in antibody reagent consumption, which is one of the most expensive components of flow cytometry that limit high throughput applications. Taking into account barcoding dye costs, we estimate a 41-fold fold decrease in costs using FCB in comparison to no barcoding flow cytometry for a 96-well plate. Initial analysis of data generated using this FCB protocol typically requires about 40 min, an amount of time for a validated assay and panel. This amount of analysis time is comparable to that needed for the analysis of samples collected without barcoding.

11.3 FCB increases data robustness

In a typical flow cytometric workflow, each individual sample must be fixed, permeabilized, stained, washed, and cytometrically analyzed individually. This introduces a high degree of variability due to pipetting error, inconsistent staining, as well as other sources of liquid handling error. The ability of the barcoding workflow to pool samples prior to functional staining enables uniform staining of samples and allows for more robust comparisons between well conditions. Additionally, automatically debarcoding data using DebarcodeR, eliminates operator bias and error introduced in manual debarcoding, increasing the reproducibility of data and allowing for more reliable comparisons between experiments with multiple internal positive and negative controls.

11.4 Modular assay composition

As previously outlined, the protocol is modular and readily adjusted according to the goal of the experiment as summarized in Fig. 1. The ability to perform experiments using established cell lines or heterogeneous cell mixtures such as PBMCs and single cell suspensions of human tumors without having to make significant protocol changes is particularly useful. When this feature is combined with multiple staining modules that can be swapped out entirely, or mixed and matched to fit one's needs. This assay thus enables a range of immune-oncology discovery studies, with applicability from cell biology to natural product drug discovery.

11.5 High-throughput assay multiplexing screening tool

High throughput bioactive small molecule discovery efforts typically rely on single functional assays, such as in vitro or in vivo biochemical assays, or phenotypic assays, such as cytotoxicity assays. Because single functional readouts generally do not capture comprehensive changes in cell states or identities, it is useful to develop modules that capture a wide range of possible cellular activities by multiple activity assays. Focusing on one functional assay means it is unknown if small molecule library hits generated will be selective agents and using biochemical assays or cell lines does not account for the heterogeneity of cell types present in physiologically relevant environments. Phenotypic screening with cytotoxicity assays also possesses limitations. For example, cell death as an endpoint provides limited molecular insights and is not typically interpretable in heterogenous primary cell samples. This protocol enables high-throughput screening with single cell resolution and multiplexed functional readouts to generate high dimensional activity profiles. This enables not only identification of perturbagenic active compounds, but also gaining insight into the molecular basis for the observed activity, for example induction of regulated cell death pathways. Additionally, the single cell resolution enables observation of activity profiles at the sample level as well as within the discrete cell population in a mixed cell type sample, such as cells in the G_1 vs G_2 cell cycle phase or monocytes vs. leukemia blasts.

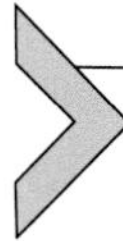

12. Limitations

12.1 Assay throughput

FCB provides substantially higher throughput than non-barcoded samples but is limited by the barcoding dye spectral separation required for accurate well assignment. Throughput is also limited by spectral overlap constraints that need to be taken into account when balancing module composition with barcoding dyes. Barcoding could potentially be expanded to include more wells (e.g., 96-well barcode), but this would require the sacrifice of a cell function or identity antibody to prevent spectral overlap. By performing two 48-well barcode experiments with 6 readouts and 16 total control wells (8 per barcode) on a 96 well plate, there are more experimental individual immunoassays than performing a 96-well barcode experiment with 8 additional experimental wells and 1 less functional readout (96 wells–16 controls) $\times$ 6 functional readouts = 480 immunoassays vs (96 wells–8 controls) $\times$ 5 functional readouts = 440 immunoassays.

12.2 Small molecule inputs

This procedure may be complicated by highly fluorescent or highly cytotoxic compounds. Low levels of compound fluorescence are tolerated and false positives due to spectral overlap with module fluorescent antibodies are easily identified by shifts of the red band tracking with shifts of the black band (Fig. 3). For example, most polyaromatic secondary metabolites such as anthracyclines do not interfere with FCB. However, excessive fluorescence overlapping too much with barcoding dye emissions can cause barcoding levels to "bleed" into each other, making debarcoding manually or using DebarcodeR very challenging or sometimes impossible. Additionally, while discovery of highly cytotoxic molecules may be of interest, these cells will be removed when gating for viability during processing and no functional readouts will be observed. This may require optimization of the small molecule input, (e.g. concentration, incubation time) in order to generate a maximal observable shift in functional marker changes.

12.3 Optimization

Established methods may require optimization when other variables are changed. This is especially true when introducing new cell types or introducing new fluorescent antibodies. Different cell types have different overall protein levels which may require optimization of barcode dye levels when attempting to barcode a new cell input. Additionally, cell size, basal core cell functions, signal pathway activation, surface marker prevalence, and more are highly variable between different cell input so optimization of established modules may also be required when introducing a new cell input. Most importantly, when introducing new fluorescent antibodies to a module or developing a new module, new antibodies must be titrated for optimal staining intensity while preventing excessive compensation issues.

13. Optimization and troubleshooting

13.1 Barcoding quality

One of the most common issues to occur in this type of protocol is for the barcoding to fail. When the barcoding fails it is often clearly seen at the cytometer or in basic analysis. In an ideal experiment, by plotting PB vs PO, a tidy square entirely on scale with clean diagonal lines or even 48 identifiable clusters should be observed (see Fig. 2). When plotting PB vs AlexaFluor 750, 8 distinct populations should be distinguishable and when plotting PO vs AlexaFluor 750, six distinct groups should be distinguishable. When barcoding fails, commonly encountered phenomena include indistinguishable smeared PO or PB levels, multiple barcoding grids, and a messier plot that is hard to distinguish. Before attempting the full protocol, we recommend completing a "barcoding only" test of untreated cells, especially when attempting to barcode a new cell line. The most commonly encountered issues effecting barcoding are dead and stressed cells, cell clumping, off scale cells (see below), or missing wells.

13.2 Dead or stressed cells

The most common cause of dead and stressed cells as well as cell clumping is improper cell preparation. Cells need to be maintained and passaged properly to ensure successful barcoding. Improperly cultured cells (starved cells, cells at too high of density, cells at too low of density, low media pH) can impact barcoding dye uptake as well as cellular function, skewing baseline and experimental marker profiles. Additionally, single cell suspension needs to be maintained throughout the protocol for the barcoding to work correctly. Failure to maintain single cell suspension may result in the formation of cytometric artifacts such as doublets and clumps that take up inconsistent levels of dye. A common issue with difficult cell types is achieving single cell suspension without killing or damaging cells. If working with a difficult cell type, it is advisable to evaluate different dissociation techniques or a waiting period between dissociation and barcoding. The AlexaFluor 700 viability stain will help quickly determine if cells are alive and/or clumping; a quick test if you are having issues would be to try different dissociation preparation techniques, stain each condition with AlexaFluor 700 only, and run on the cytometer to optimize before proceeding to do a complete FCB test.

13.3 Off scale cells and missing wells

In cytometry plots, a cell is considered 'off scale' when the measurement intensity exceeds the maximum for that instrument and detection channel and the cells appear at the very edge of a plot (e.g., on the right of the x-axis). If a large number of cells are off scale, or if entire wells are missing after debarcoding, it is possible that there is a barcoding or staining issue. Other options include additional fluorescence from compounds or a large amount of activity from a compound either killing all the cells or significantly changing their size or fluorescence properties. For off scale troubleshooting resulting from barcoding issues (i.e., barcodes are off scale, as is observed when dye levels titrated for smaller blood cells are used on much larger cancer cells), it is advised to try lower concentrations of barcoding levels for PB and PO. Cell size affects dye uptake, so if working with large cells, barcode dye concentrations may need to be diluted. Be sure to maintain dilution spacing between barcode levels to maintain separation. If wells are missing post debarcoding, then the most likely issue is a cell transferring issue. If cell pellets at any stage of the protocol are not sufficiently dispersed and resuspended, cells may fail to be carried over. Make sure cell pellets are vortexed thoroughly and pipetted up and down repeatedly to resuspend the pellet, especially when moving between plates.

14. Conclusion

This chapter aims to enable multiplexable high-content data production using barcoded flow cytometry for drug discovery and aims to balance cellular throughput, data robustness, and economy. As the assay is compatible with diverse chemistries, cell targets, and cytometry modules, it allows for immuno-oncological investigation across a range of topic areas, including secondary metabolite discovery in primary human cancer. Ultimately, we hope this protocol will remove that barrier of entry to use of multiplexed cell-based assays in discovery settings and will encourage collaborations between flow cytometry, chemical biology, and drug discovery communities.

Acknowledgments

The authors thank the National Institutes of Health for research ([R01 GM092218, Brian O. Bachmann], R01 CA226833 [Brian O. Bachmann, Jonathan M. Irish], U01 TR002625 [Jonathan M. Irish, Brian O. Bachmann]), training support (T32 GM007347 [Henry A.M. Schares]), and support of cores via the Vanderbilt-Ingram Cancer Center (P30 CA68485). The authors are grateful to Nalin Leelatian and David Earl for their initial efforts at developing FCB protocols, to Benjamin J. Reisman for developing DebarcodeR and its associated protocols, and to Rebecca Ihrie and Laura Geben for the use of example mouse cell data. Flow cytometry experiments were performed in the VUMC Flow Cytometry Shared Resource.

Glossary

COMPOUND_PLATE 96-Well tissue culture plate with either a single set of 48 small molecule inputs or two separate sets 48 small molecule inputs. Typically, the wells include 40 test conditions (e.g., small molecules, secondary metabolites, doses, timepoints; see Fig. 1) and 8 controls. For example, staurosporine is a control compound typically used to reliably activate cleaved-Caspase 3 (c-CAS3).

TREATED_PLATE 96-Well Tissue Culture Plate that started as the **COMPOUND_PLATE** but is renamed after cells are added to reflect experimental conditions.

FIX_PLATE Falcon 96-well V-bottom polypropylene microplate containing PFA where cells from the compound plate are transferred for fixation and permeabilization.

BARCODE_PLATE Falcon 96-well V-bottom polypropylene microplate prepared with dyes of PO, PB, and Ax750. This plate can be stored sealed at −80 °C. Cells will be transferred into this plate for fluorescent cell barcoding (FCB).

UNSTAINED_CELLS Unstained cells used to set up fluorescent cytometer and to make some compensation controls.

STAINED_CELLS Cells that are fully barcoded and stained with antibodies and any other readouts; this sample represents the main experimental sample of the assay (example data shown in Fig. 2).

Ax700_CELLS cells from **FIX_PLATE** that don't continue to barcoding and have been treated with compound and stained with Ax700 only (no additional stains) when collected. A control used to check for background fluorescence from cells or compounds in troubleshooting.

UTC_CELLS Cells from **Ax700_CELLS** tube that continue to get stained, but only with the highest levels of PO, PB, and Ax750; this control represents the highest level of each dye from barcoding and is used in debarcoding and troubleshooting.

FCB_CELLS Full set of cells from all wells of **BARCODE_PLATE** that do not continue to get dyed with antibody stains and become a control during debarcoding and in the resulting output of debarcodeR (baseline fluorescence represented by red cells in Fig. 3).

COMP_TUBE_[stain] One each per readout (e.g., an antibody channel). Compensation tubes are made fresh for each experiment and used as controls in flow cytometry. These will include controls for all barcoding dyes and reagents used to probe cellular functions (antibodies and dyes). For each **COMP_TUBE**, the [stain] section will denote what the stain is, such as **COMP_TUBE_PE** for the PE-conjugated c-CAS3 antibody captured by beads.

STAINING_MIX A large batch of pre-mixed antibodies that will be aliquoted out and applied to multiple tubes of cells during antibody staining.

References

Bodenmiller, B., Zunder, E. R., Finck, R., Chen, T. J., Savig, E. S., Bruggner, R. V., et al. (2012). Multiplexed mass cytometry profiling of cellular states perturbed by small-molecule regulators. *Nature Biotechnology*, *30*(9), 858–867. https://doi.org/10.1038/nbt.2317.

Boyce, J. H., Reisman, B. J., Bachmann, B. O., & Porco, J. A., Jr. (2021). Synthesis and multiplexed activity profiling of synthetic acylphloroglucinol scaffolds. *Angewandte Chemie (International Ed. in English)*, *60*(3), 1263–1272. https://doi.org/10.1002/anie.202010338.

Brodin, P., Jojic, V., Gao, T., Bhattacharya, S., Angel, C. J., Furman, D., et al. (2015). Variation in the human immune system is largely driven by non-heritable influences. *Cell*, *160*(1–2), 37–47. https://doi.org/10.1016/j.cell.2014.12.020.

D'Antonio, M., Woodruff, G., Nathanson, J. L., D'Antonio-Chronowska, A., Arias, A., Matsui, H., et al. (2017). High-throughput and cost-effective characterization of induced pluripotent stem cells. *Stem Cell Reports*, *8*(4), 1101–1111. https://doi.org/10.1016/j.stemcr.2017.03.011.

Earl, D. C., Ferrell, P. B., Jr., Leelatian, N., Froese, J. T., Reisman, B. J., Irish, J. M., et al. (2018). Discovery of human cell selective effector molecules using single cell multiplexed activity metabolomics. *Nature Communications*, *9*(1), 39. https://doi.org/10.1038/s41467-017-02470-8.

Fucikova, J., Kepp, O., Kasikova, L., Petroni, G., Yamazaki, T., Liu, P., et al. (2020). Detection of immunogenic cell death and its relevance for cancer therapy. *Cell Death & Disease*, *11*(11), 1013. https://doi.org/10.1038/s41419-020-03221-2.

Galluzzi, L., Buque, A., Kepp, O., Zitvogel, L., & Kroemer, G. (2017). Immunogenic cell death in cancer and infectious disease. *Nature Reviews. Immunology*, *17*(2), 97–111. https://doi.org/10.1038/nri.2016.107.

Giudice, V., Feng, X., Kajigaya, S., Young, N. S., & Biancotto, A. (2017). Optimization and standardization of fluorescent cell barcoding for multiplexed flow cytometric phenotyping. *Cytometry. Part A*, *91*(7), 694–703. https://doi.org/10.1002/cyto.a.23162.

Irish, J. M., & Doxie, D. B. (2014). High-dimensional single-cell cancer biology. *Current Topics in Microbiology and Immunology*, *377*, 1–21. https://doi.org/10.1007/82_2014_367.

Irish, J. M., Myklebust, J. H., Alizadeh, A. A., Houot, R., Sharman, J. P., Czerwinski, D. K., et al. (2010). B-cell signaling networks reveal a negative prognostic human lymphoma cell subset that emerges during tumor progression. *Proceedings of the National Academy of Sciences of the United States of America*, *107*(29), 12747–12754. https://doi.org/10.1073/pnas.1002057107.

Krutzik, P. O., & Nolan, G. P. (2006). Fluorescent cell barcoding in flow cytometry allows high-throughput drug screening and signaling profiling. *Nature Methods, 3*(5), 361–368. https://doi.org/10.1038/nmeth872.

Krutzik, P. O., Clutter, M. R., Trejo, A., & Nolan, G. P. (2011). Fluorescent cell barcoding for multiplex flow cytometry. *Current Protocols in Cytometry*. Chapter 6, Unit 6 31 https://doi.org/10.1002/0471142956.cy0631s55.

Krutzik, P. O., Crane, J. M., Clutter, M. R., & Nolan, G. P. (2008). High-content single-cell drug screening with phosphospecific flow cytometry. *Nature Chemical Biology, 4*(2), 132–142. https://doi.org/10.1038/nchembio.2007.59.

Krutzik, P. O., Irish, J. M., Nolan, G. P., & Perez, O. D. (2004). Analysis of protein phosphorylation and cellular signaling events by flow cytometry: Techniques and clinical applications. *Clinical Immunology, 110*(3), 206–221. https://doi.org/10.1016/j.clim.2003.11.009.

Leelatian, N., Doxie, D. B., Greenplate, A. R., Mobley, B. C., Lehman, J. M., Sinnaeve, J., et al. (2017). Single cell analysis of human tissues and solid tumors with mass cytometry. *Cytometry. Part B, Clinical Cytometry, 92*(1), 68–78. https://doi.org/10.1002/cyto.b.21481.

Manohar, S., Shah, P., Biswas, S., Mukadam, A., Joshi, M., & Viswanathan, G. (2019). Combining fluorescent cell barcoding and flow cytometry-based phospho-ERK1/2 detection at short time scales in adherent cells. *Cytometry. Part A, 95*(2), 192–200. https://doi.org/10.1002/cyto.a.23602.

Menard, C., Martin, F., Apetoh, L., Bouyer, F., & Ghiringhelli, F. (2008). Cancer chemotherapy: Not only a direct cytotoxic effect, but also an adjuvant for antitumor immunity. *Cancer Immunology, Immunotherapy, 57*(11), 1579–1587. https://doi.org/10.1007/s00262-008-0505-6.

Michaud, M., Martins, I., Sukkurwala, A. Q., Adjemian, S., Ma, Y., Pellegatti, P., et al. (2011). Autophagy-dependent anticancer immune responses induced by chemotherapeutic agents in mice. *Science, 334*(6062), 1573–1577. https://doi.org/10.1126/science.1208347.

Reisman, B. J., Barone, S. M., Bachmann, B. O., & Irish, J. M. (2021). DebarcodeR increases fluorescent cell barcoding capacity and accuracy. *Cytometry. Part A, 99*(9), 946–953. https://doi.org/10.1002/cyto.a.24363.

Roederer, M. (2002). Compensation in flow cytometry. *Current Protocols in Cytometry*. Chapter 1, Unit 1 14 https://doi.org/10.1002/0471142956.cy0114s22.

Sriram, G., Milling, L. E., Chen, J. K., Kong, Y. W., Joughin, B. A., Abraham, W., et al. (2021). The injury response to DNA damage in live tumor cells promotes antitumor immunity. *Science Signaling, 14*(705), eabc4764. https://doi.org/10.1126/scisignal.abc4764.

Further reading

Balsamo, J. A., Penton, K. E., Zhao, Z., Hayes, M. J., Lima, S. M., Irish, J. M., & Bachmann, B. O. (2022). An immunogenic cell injury module for the single-cell multiplexed activity metabolomics platform to identify promising anti-cancer natural products. *The Journal of Biological Chemistry, 298*(9), 102300. https://doi.org/10.1016/j.jbc.2022.102300.

Printed in the United States
by Baker & Taylor Publisher Services